WALNUT CREEK

D0825435

THE STORY OF RATS

THE STORY OF RATS

Their impact on us,
and our impact on them

S. Anthony Barnett

WITHDRAWN

CONTRA COSTA COUNTY LIBRARY

ALLEN&UNWIN

3 1901 03018 0121

First published in 2001

Copyright © S. Anthony Barnett 2001

All rights reserved. No part of this book may be reproduced
or transmitted in any form or by any means, electronic or
mechanical, including photocopying, recording or by any
information storage and retrieval system, without prior
permission in writing from the publisher. The *Australian
Copyright Act 1968* (the Act) allows a maximum of one chapter
or 10% of this book, whichever is the greater, to be photocopied
by an educational institution for its educational purposes
provided that the educational institution (or body that
administers it) has given a remuneration notice to Copyright
Agency Limited (CAL) under the Act.

Allen & Unwin
83 Alexander Street
Crows Nest NSW 2065
Australia
Phone: (61 2) 8425 0100
Fax: (61 2) 9906 2218
Email: info@allenandunwin.com
Web: www.allenandunwin.com

National Library of Australia
Cataloguing-in-Publication entry:

Barnett, S. A. (Samuel Anthony), 1915– .
    The story of rats: their impact on us and our impact on
    them.

    Bibliography.
    Includes index.
    ISBN 1 86508 519 7.

    1. Rats. 2. Rats as carriers of disease. 3. Rats—
    Behavior. 4. Rats as laboratory animals. I. Title.

599.352

Set in 11/14.5 pt Garamond 3 by Bookhouse, Sydney
Printed by Griffin Press, Adelaide
Text design by Simon Paterson

10 9 8 7 6 5 4 3 2 1

Mr Speaker, I smell a rat; I see him forming in the air and darkening the sky; but I'll nip him in the bud.

Sir Boyle Roche, eighteenth-century English MP

CONTENTS

Illustrations   x
Preface   xii

PART I HISTORIES   1

1 Tales of Rats   3
   Abominations and Horrors   5
   Magic, Sport and Nourishment   8

2 Naming and Taming   14
   The Names of the Species   14
   Conflict Between Species   17
   Taming and Domestication   21

3 All Fall Down   24
   Mud Fever   24
   How Not to Handle Wild Rats   26
   Plague   27
   The Black Death   29
   Public Health   36
   Yersinia pestis   39
   Modern Plague   41
   New Viruses   44

PART II QUESTIONS AND ANSWERS   51

4   A Battle of Wits?   53

     What Do Rats Really Do?   54
     Neophobia: Can Curiosity Kill the Rat?   56
     Neophilia and Natural Selection   58

5   Do Rats Think?   64

     Finding the Way   65
     Students in Mazes: A Detour   67
     Patrolling and Sampling   68
     Mechanistic Psychology   71
     Do Rats Get Bored?   77
     Playing to Learn   78
     The Question of Thinking   81
     Are People Ever Like Rats?   83
     The Tree of Knowledge   85
     Thought About Thought   88

6   Are Rats Gluttons?   91

     Selecting from a Menu   92
     Consuming Passions   97
     Poison Shyness   98
     Psychologists Develop an Aversion   101
     A Clinical Contribution   103
     Social Feeding: Black Rats in Pine Forests   105
     Social Feeding: White Rats in Cages   106
     Early Learning   111
     Learning: Limited and Unlimited   113

Contents

7 All in Their Genes?   115
   Some Unnatural Selection   116
   Petted Rats   119
   Are Rats Darwinian?   122

8 Rat Societies   128
   Social Signals: The Show of Violence   129
   Close Encounters   136
   Assault with Battery   137
   Rat Slaughter   139
   A Diversity of Species   142

PART III THE BLINDNESS OF RESEARCH   147

9 Population Explosions   149
   Density Dependence   151
   Underground Warfare   153
   Awkward Questions   156
   Rice Rats and the Economic Threshold   158
   Social Sedation versus Social Stress   159

10 Social Life and Death   164
   The Search for Instability   165
   More Social Stress   167
   Is DUO Darwinian?   171

11 Nature and Human Nature   173

Glossary   177
Notes on sources   186
References   188
Index   196

ILLUSTRATIONS

Page

4   Rodents in ancient Egypt

7   Giant mediaeval rat seizes cat

9   Hindu deity with rat

15   This 'brown' rat is black

17   The Indian mole rat

19   Imagined conflict between 'black' and 'brown' rats

22   White rat

27   Protective clothing

32   A rat catcher in Shakespeare's time

37   Protection against plague

55   Sugar cane in Hawaii attacked by rats

59   An automated residential maze

66   Vibrissae

69   Sniffing the air

70   Hooded rate in Y-maze

72   A Pavlovian dog

75   Rat in Skinner box

81   Training to jump

92   Handling a wheat grain

93   Rat chooses from offered foods

107 Young cone stripper

107 Cone stripping

132 Crawling under

132 Grooming

134 'Threat posture'

135 Attack

135 Sniffing body odours

136 'Boxing'

141 Tree shrew with raised hairs

150 Maternal behaviour

155 Research in sewer

160 Social sedation

162 Mole rats in Calcutta godown

169 Black rats expending energy

PREFACE

Can it be—it must be—that you are that
embodiment of the incorporeal, that elusive
yet ineluctable being to whom through the
generations novelists have so unavailingly made
invocation; in short, the *Gentle Reader?*

HENRY JAMES

When my publisher, Ian Bowring, received the manuscript
of *Science, Myth or Magic?*, he suggested that, after half
a century of research on rodents and their impacts on people,
I might have something useful to say about them. So I asked
several thoughtful persons what they would like to find in a
short book on science and rats. One, a true friend, warned me
against being boring and worthy. Another said that the title
I proposed, *Wild Rats*, suggested a thriller. At first I disregarded
that; but later I realised that passages in my narrative, especially
those on death of unknown origin, have something in common
with a detective story. A third said that she would look for
likenesses of rats to herself. My immediate response was, 'And
unlikenesses?', which led to an energetic but still unfinished
discussion. Lastly, one suggested a social history of human

relationships with rats; and this is a large part of what I have attempted.

The possibility of likening lowly mammals to human beings has led to many researches, especially on tame varieties, in laboratories. It is said that inside every fat man is a thin one trying to get out. A case exists for saying that inside every white rat, too, is a wild one. Despite my primary interest in the wild forms, I write a good deal about these strange creatures. Several chapters outline findings which make us ask whether animals in laboratories tell us anything useful about our own social lives.

Still more important, in the twenty-first century famine and epidemics remain with us. We are continually struggling to feed ourselves, to avoid infection and, above all, to preserve the biosphere for our descendants. Simple solutions are few. And so, in this book, we meet not only scientific methods and achievements which help us to solve our problems, but also the need for more knowledge and more effective action.

PART I
HISTORIES

Human attitudes to rats depend on custom and circumstance: they include terror at the threats they represent to our food and health and welcome as a source of nourishment; and they range from disgust to curiosity.

1
TALES OF RATS

'Matilda Briggs', said Holmes, 'was ... a ship
which is associated with the giant rat of
Sumatra, a story for which the world is not
yet prepared.'

A. CONAN DOYLE,
The Adventure of the Sussex Vampire

Early in the fifteenth century, the Bishop of Autun in
France put rats under a formal curse. He was following
the example of much earlier priests and gurus, notably the
authors of the ancient Indian scriptures, the *Atharva Veda*, who
exhorted a supernatural being:

O, Ashwini. Kill the burrowing rodents which devastate
our food grains, slice their hearts, break their necks, plug
their mouths, so that they cannot destroy our food.

Others, similarly harassed, have been more civil. In Greece,
a farmer attacked by mice or rats (they were not clearly
distinguished) was told to write on a sheet of paper as follows:

I adjure you, ye mice here present, that ye neither injure
me, nor suffer another mouse to do so. I give you yonder

Rats (and mice) were a prominent feature of daily life in the ancient world. Here rodents (species uncertain) are besieging a fortress occupied by cats. The drawing, from around 1200 BC, may be a satirical comment on a pharaoh's triumph in war. (After Jaromir Malek 1993, The Cat in Ancient Egypt, *British Museum)*

field. But if I catch you here again, by the mother of the gods, I will rend you in seven pieces.

The anthropologist James Frazer (1854–1941), in *The Golden Bough*, describes how an even more courteous American farmer wrote a letter to his rats, in which he said that crops were short, that he could not afford to keep rats during the winter, that he had been very kind to them but, for their own good, they should go to some of his neighbours who had more grain. 'This document he pinned to a post in his barn for the rats to read.'

Occasionally, authors have gone still further and written approvingly of rats' morals. Here is a once popular example from Thomas Bell's *A History of British Quadrupeds*, published in 1837.

Although its disposition appears to be naturally exceedingly ferocious... Mr Jesse gives us an anecdote [of a rat] exhibiting a degree of tenderness and care towards the disabled and aged members of their community, which, were it imitated by Christian men, would either render our Poor Laws unnecessary, or remove the opprobrium which their maladministration too often causes to attach to them. His informant, the Rev. Mr. Ferryman, walking out in some meadows one evening, 'observed a great number of rats in the act of migrating from one place to another ... His astonishment was great when he saw an old blind Rat, which held a piece of stick at one end in its mouth, while another Rat had hold of the other end of it, and thus conducted his blind companion.'

ABOMINATIONS AND HORRORS

On the whole, however, approval of wild rats is rare: they are more often greeted with horror, disgust or vigorous pursuit. In the Bible, in passages written nearly three millennia ago, the children of Israel are told:

> These shall be unclean unto you among the things that creep upon the earth; the weasel and the mouse and the tortoise after his kind (*Leviticus* 11, 29).

Here 'mouse' again means, or includes, 'rat'; for until recently writers rarely distinguished rats from mice.

The number of the biblical 'abominations' is large and the selection of species to be rejected has caused much puzzlement. But, as the anthropologist Mary Douglas points out in *Purity and Danger*, all are presented as unclean, anomalous, unholy and therefore unacceptable to the faithful. As late as the nineteenth century, even a textbook of zoology describes rats as 'hateful and rapacious'.

Etymology

In Latin the word for 'rat' is *mus*; in classical Greek it is σμίνθος (sminthos). The god Apollo, among his many duties, had the task of destroying rats and mice: the temple of Apollo Smintheus, in the city of Chrysa, had a sculpture of a rodent at the foot of the god.

According to Samuel Johnson's *Dictionary of the English Language* (1755), the Italian word for rat can be derived from the Latin *mus, muris*, a mouse, by the following process: *murus, muratus, ratus, rato, ratto*; but the German *ratz* is a more 'natural' etymology. After all this, Johnson's definition of 'rat' is an anticlimax: 'An animal of the mouse kind that infests houses and ships'.

In an edition of the *Oxford English Dictionary* published two centuries later, 'rat' is admitted to have an unknown origin. The primary definition is badly out of date: 'any rodent of certain of the larger species of the genus *Mus*'. But rat is also an 'opprobrious epithet' used in politics for one who deserts his party or for a worker who refuses to strike with others.

This tradition has greatly helped some modern writers of fiction. In George Orwell's *Nineteen Eighty-Four*, a famous tract against tyranny and 'thought police', the hero is threatened with a cage holding 'enormous rats'. The torturer explains that in the poor quarters of the town 'a woman dare not leave her baby alone... even for five minutes. The rats are certain to attack it. Within quite a small time they will strip it to the bones.'

After uttering this fearsome improbability, the torturer describes how rats 'show astonishing intelligence in knowing when a human being is helpless'; and he explains that the

In Europe in the late Middle Ages rats were sometimes held to be supernatural creatures and portents of evil. A giant rat seizes a cat. (After Beryl Rowland 1974, Animals with Human Faces, *Allen & Unwin)*

cage will be fitted over the hero's head, the rats will be released and at once bore into his head and probably attack his eyes first. This hideous fantasy has no connection with what rats would actually do (they would probably be terrified); but the writing is so skilful that even today it remains effective.

Lament for a Maker, a distinguished crime story by Michael Innes (a former professor of English) written a few years before *Nineteen Eighty-Four*, also uses rats to create atmosphere; but it includes an element of parody of the 'Gothic' romances popular in England in the eighteenth and nineteenth centuries and satirised by Jane Austen in *Northanger Abbey*. Much of the action occurs in a crumbling, isolated Scottish castle with a large rat population, 'Mongst horrid shapes, and shrieks, and sights unholy'.

We are told that the people in the castle ate mainly from

tins. A zoologist is therefore tempted to ask how all those rats managed to feed themselves. That, however, would be unkind and inappropriate. In the story, moreover, the rats carry the narrative forward: some are (quite plausibly) tamed and (less plausibly) employed by a prisoner to convey scrawled messages, fastened to their legs with cotton threads.

MAGIC, SPORT AND NOURISHMENT

Despite the ancient Indian objurgation above, Hinduism is not consistently hostile to rats. A popular deity, the rotund, four-armed Ganesa, who has an elephant's head, is often represented as accompanied by a rat or even riding on one. Ganesa is a god of literacy and learning, who (bless him) insists that everything written should be readily understood. His followers commonly seek his support at the beginning of an undertaking such as producing a book. Perhaps I should have propitiated him before I began to write this one. But I am not quite clear why he is so closely associated with rats.

For other peoples, rats have or had magical powers which are more readily understood. South African warriors twisted tufts of rats' hair in their own hair: this, they said, conferred on them the agility of a rat, and so helped them to avoid an enemy's spear.

Teeth, appropriately, loom large in some rodent magic. In Germany it was held that a child's milk tooth should be thrown backwards over one's head, accompanied by an incantation demanding that rats should provide the child

Ganesa, a leading Hindu deity with an elephant's head, is propitiated at the beginning of any kind of undertaking. He regularly rides on, or is accompanied by, a rat. (After Edward Tyomkin 1994, The Hindu Pantheon, *Garnet Publishing)*

with an 'iron' tooth. In Raratonga a child's milk tooth was cast into a hut's thatch, where rats lived, in exchange for a new and more powerful dentition.

Rats can also provide sport. In the Friendly Islands (Fiji) a group of Tongans are described as hunting rats for pleasure—a counterpart of the shooting of birds or hunting of foxes practised by others. Each of two chiefs assembles a party equipped with bows and arrows. The arrows are very long and beautifully made but have only a short range. A path is marked by two attendants who spit out fragments of roasted coconut on each side. Nobody, except the hunting parties, may now use the path. The hunters walk in single file, with members of the rival teams alternating and all imitating rat squeaks. The rats, which seem to be remarkably helpful, then emerge from their burrows. The winning party is the first to shoot ten rats. According to the anthropologist J.G. Wood, birds of any kind are counted as rats.

Not all rat hunting has been merely sport. In ancient Japan, artists prized rats' whiskers for their brushes, which made possible the delicacy of traditional painting. More often, rats have been valued for their skins. In England, in the fourteenth century, a skin cost one farthing (then a useful sum). During the Second World War, a kind colleague produced rat leather, made from wild *Rattus norvegicus* trapped for research, and gave me some. Though thin, it made useful elbow patches.

In sailing ships, in which rats were as familiar as bed bugs, pursuing them no doubt relieved boredom; but they were also eaten, which suggests an interesting possibility. Famously, sailors on long voyages were liable to scurvy, owing—we now know—to lack of vitamin C (ascorbic acid). The rats, like the crew, lived largely on salt pork, dried peas and ship's biscuit; but they were better off than the sailors for, like most mammals, they can synthesise the vitamin from other food substances. Perhaps, on long voyages, raw or lightly cooked rats were a useful source of vitamin C and so helped to prevent scurvy. A survivor of one of the appalling voyages in the south seas of Fernando Magellan (?1480–1521) describes living on powdered, wormy biscuit stinking of rat urine; and he complains that he and his companions could not get enough rats to eat.

In some parts of India, rats are still an important dietary supplement. The rice fields in Karnatica, in the south, are small and lined with banks (or bunds) which house fabulous numbers of rodents. A single field may harbour the Indian gerbil (*Tatera indica*)—a creature of charming appearance which is an absolute menace to farmers; the Indian mole rat (*Bandicota bengalensis,* page 17)—which is probably even worse, and perhaps the soft-furred rat (*Rattus meltada*).

In the tops of the banks, a close look also reveals the small holes made by mice, of several species closely related to the house mouse: they are preyed on by the larger rodents and

Strictly for Gourmets

A reader who wishes to eat rats can resort to a standard work, *Guide to Good Food and Wines*, by André Simon. For both Norways and black rats he recommends a stuffing made from breadcrumbs, the minced liver and heart of the rat and sweet herbs, pepper and salt. Roast for a few minutes in a hot oven. Young rats, he adds, may be made into pies.

Larousse Gastronomique has an alternative, derived from the practice of coopers in the wine stores of the Gironde. After they had been skinned and cleaned, the rats were seasoned with oil and plenty of shallots and grilled over an open fire.

Both authorities emphasise the excellence of rat flesh; but, I am sorry, neither gives the number of rats needed per person.

seem to live rather precarious lives. If, however, most of the larger species are killed, they quickly multiply.

The villages have rodents of different species. They include house or roof rats (*Rattus rattus*) and, in smaller numbers, a true giant, the large bandicoot rat (*Bandicota indica*), which can weigh over 1000 grams. This animal is rather sluggish and often merely grunts like a pig when disturbed; hence its name in Telugu is *pandi-koku* (the word from which 'bandicoot' is derived).

All these rodents depend almost entirely on food grown by the villagers. Seeds may be eaten as soon as they are sown; early shoots are nibbled; ripe grain is cut down and eaten or stored in burrows; harvested grain is consumed in the threshing yard and in the village stores.

Above I quote an ancient Hindu prayer (or curse). Most of the farmers are Hindus for whom, as for the Israelis, 'vermin' are unclean, not to be touched and emphatically not to be eaten. But many regions of India also have unobtrusive inhabitants who are not Hindus of any kind —or Muslim or Christian. These are the tribals.

Those I met are small, dark, quiet people, called Irulas, with very hard hands. At harvest time they are hired by farmers to come, in teams of three or four, to catch the rats and gerbils: they smoke them out or dig up the burrows and catch the animals in nets. They are very skilled and the result is often an impressive mound of dead rodents. Romulus Whitaker, the President of the Irulas Snake-Catchers Cooperative, told me that sixty to seventy Irulas can catch 10 000 to 12 000 rodents in a month. They cook and eat the bodies. (I have never tried them; but Whitaker says that some species are very palatable.) In this way many rodents are killed before they reach the threshing grounds or the winter stores; and the tribals get some first class protein.

As this chapter shows, our relationships with rats reflect human cultural diversity; but, so far, we have not met the aspect of human action which is central to this book: science. That begins to emerge in the next chapter.

2
NAMING AND TAMING

'When I use a word,' Humpty Dumpty said,
in rather a scornful tone, 'it means just what
I choose it to mean—neither more nor less.'
　　'The question is,' said Alice, 'whether you
can make words mean so many different things.'
　　'The question is,' said Humpty Dumpty,
'which is to be master—that's all.'
　　　　LEWIS CARROLL, *Through the Looking Glass*

We now turn to another kind of magic, that of words. In some human groups, knowing a person's name is held to confer power over that person. In biology, precise description and agreed naming also give power. Reliable identification is often essential if one wishes to preserve or to breed a species; or to destroy it. Classifying and naming are therefore of vast importance.

THE NAMES OF THE SPECIES

The colloquial names of the two most widespread species of rat reflect their omnipresence and are notoriously confusing.

14

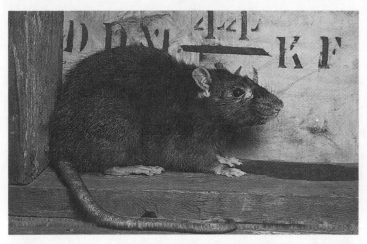

A black (melanic) 'brown' rat, Rattus norvegicus. *(Courtesy John Markham)*

The 'common, brown (or grey), wharf, Hanoverian or Norway' rat (*Rattus norvegicus*) is, in many parts of the world, especially the tropics, not common at all; it is normally grey-brown but sometimes black; in its domestic forms it is usually white, or white and black; it is only incidentally associated with wharves; and it does not come from Hanover or Norway. In the early days of modern Latin names it was called *Mus norvegicus*, which seems to signify a kind of mouse. Later, it was named *Epimys* instead of *Mus*, which suggests a superior mouse. In the rest of this book, in view of its modern Latin name (finally determined by the rules of biological nomenclature), I reluctantly refer to it as the Norway rat.

The 'old English, black (or blue), roof, alexandrine or ship rat' (*Rattus rattus*) is not English by origin; it is often tawny with a white belly, less often black; it can (rarely) be white.

It is commonly found in surroundings quite different from those of a roof and, today, not often in ships. It has been said to exist in several subspecies distinguished by their colour: black (called *Rattus rattus rattus*), brown (*R. r. alexandrinus*) and tawny (*R. r. frugivorus*). But these colour 'morphs' should not be treated as subspecies: they appear in all parts of the species' vast range and may all occur among the offspring of a single female. The light coloured forms are evidently typical: the black form is common only in the northern parts of its distribution, in Europe and North America.

Like Norways, individuals if taken young can be tamed (I have done it); but *R. rattus* has never been systematically domesticated. With some hesitation, and purely for convenience, in this book I call it the black rat. Until the eighteenth century, in Europe it was *the* rat. Then the Norways invaded from the East and, in northern latitudes, the black rat declined.

The names of other rats are less confusing. An example is the Indian mole rat, *Bandicota bengalensis*, which deserves special comment. It is a pest throughout India and Nepal, and in parts of Myanmar (Burma), Thailand, Sri Lanka (Ceylon), Sumatra and Java. Like the Norway, which it closely resembles, it is nocturnal and a burrower. It is numerous on farms and has been called a field or rice rat; but enormous numbers are also present in the godowns or food stores of Indian cities. In Calcutta, during the twentieth century, it mysteriously ousted the formerly flourishing Norways, just as Norways replaced black rats in Europe. It is also the dominant rat in Rangoon, the principal city of Myanmar.

This harmless looking creature, the Indian mole rat (Bandicota bengalensis), is present in uncounted millions in many Asian countries, where it is a major pest.

CONFLICT BETWEEN SPECIES

How long the black rat has been in Europe is uncertain. A skull and other remains of black rats have been reported by James Rackham from a Roman well of around 300 AD, discovered in the north of England. The well evidently served the Roman army stationed on Hadrian's Wall, the once great but now grass-grown barrier between England and Scotland. On this and other evidence, black rats had infested Britain, as they had much of mainland Europe, by the time of the Roman occupation of Britain.

For two millennia or more, the black rat seems to have been on its own (apart from the always present house mouse, *Mus domesticus*). But, in the eighteenth century, the heavier and seemingly more energetic ('aggressive') Norway invaded

Nominations

Biological naming is regulated by international bodies with strict rules which are sometimes difficult to apply. It is founded on the famous system designed by Linnaeus (1707–1778), a Swedish physician of great charm who became a professor of botany. All known species are assigned a double name, traditionally derived from Latin or Greek. Today a new name should be supported by a 'type specimen' which may be housed in a museum collection.

A species is often said to be a population of which all the individuals can interbreed and produce fertile offspring. This definition—or description—cannot be applied to all organisms, but it works well for Norways and black rats. Attempts have been made to interbreed them: sometimes success has been reported; but this seems to have been due to mistaking black Norways for black *Rattus rattus*. When (as I have seen for myself) individuals are properly identified and brought together, they do not even copulate, let alone produce young.

If two systematists describe the same species, the name published earlier is usually adopted. Norways have *norvegicus* as their species name because, in his *Outline of the Natural History of Great Britain* of 1769, J. Berkenhout used it in the first formal, Linnaean description of the species. (He mistakenly thought that the species came from Norway, but this does not invalidate the name.) The full, formal name of the species is therefore: *Rattus norvegicus* Berkenhout.

Europe from the East and, in much of Europe, largely ousted the smaller black rat.

The Norways are traditionally supposed to have succeeded by violence, like the well armed Dutch, French and British

Conflict between two species portrayed in 1883. Predatory attack by a Norway on a black rat. The reality is not quite so violent.

armies and navies of that time which crushed feebly equipped 'primitive tribes'; but the geography of the two species makes their relationship less clear. As a rule, the nearer we approach the equator we find more black rats and fewer Norways. In Mediterranean countries, light coloured black rats remain prominent. In most of Australia, which is largely subtropical or tropical, wild Norways are uncommon—which is inconvenient when one wants to do research on them; while black rats (though rarely in their black form) are easy to find. Norways in these warm climates are not violently exterminated by the rival species. Perhaps they are adversely affected by the higher temperatures.

Another peculiarity is the distribution of the two species in buildings. In London and elsewhere, early in the twentieth century, it was usual to find Norways in sewers and basements, but black rats in attics, with perhaps a disputed zone between them.

What actually happens when freeliving members of the two species meet? For a short time, in the 1940s, I was able to use some disused stables for setting up colonies of wild rats and watching them. Once they had settled down, I could lie on the floor taking photographs while the rats ran around and over me. In one experiment, 29 Norways and 19 black rats were established in adjoining enclosures. None of the blacks weighed more than 200 grams; most of the Norways were heavier. After two weeks, a barrier between the two populations was removed and a single feeding point provided at the frontier. Members of both species soon explored the newly accessible areas. Whenever a rat entered a strange nest, a disturbance followed; usually, the intruder withdrew.

A crucial event was the entry of a weighty Norway male (440 grams) into the nests of black rats: upheaval followed and culminated in flight toward the roof by the blacks. The experiment ended when, after nine days, three Norways and thirteen blacks had died. Death was not due to violence (despite the dramatic picture on page 19). Its possible causes I defer to chapter 10.

Conflict between the species is not inevitable. I have had adult males of both species living peacefully together in a large cage. If, however, in such a cage, a male of one species enters the territory of the other, it may be resisted, as in the experiment just described. But sometimes incongruous behaviour occurs: a male Norway may attempt coitus with a black male, but not a female. Perhaps male black rats have confusing odours. This is one way in which species are kept separate: the odours and other features of one are not right for the other.

We still have much to learn about what determines the relations of these species.

TAMING AND DOMESTICATION

During the nineteenth century, Norways became domesticated on a large scale and produced the various forms of laboratory rat. In 1894, at Clark University (Worcester, Massachusetts), wild Norways were adopted for research on the effects of diet and of alcoholism. This program did not last long. Perhaps the researchers were bitten too often. In 1895, white Norways were brought in and so began the use of domestic rats as models for the study of human physiology (a role partly usurped later by the house mouse, *Mus domesticus*).

In 1900, again at Clark, a psychologist, W.S. Small, founded 'white rat psychology' by launching experiments on rats in mazes. Small was inspired by a famous institution at Hampton Court, in outer London. There visitors could (and can) spend a sunny afternoon getting lost in an elaborate maze made of high, well tended hedges.

During domestication, members of a species become adapted to conditions, often including close confinement, imposed by human beings. This entails selection (planned or unplanned), because many newly captive animals are infertile. The domestic stock is descended from those, sometimes only a few, that produce young. In one early case, sixteen male and twenty female Norways were trapped, but only six of the females had litters. In 1919, these gave rise to a stock which was used in laboratories for many years.

The rat of laboratories and pet shops. Domestic Norways of a common, albino variety. (Photo by Philip Boucas, courtesy WHO)

The result of such inadvertent selection is unavoidably a population genetically different from the original wild type. Features not advantageous in freedom may now favour survival. One is a conspicuous colour, which may be welcomed by human captors but in the wild would help predators such as hawks. More important, domestic animals, when seized or restrained, rarely attack their owners, and the tendency to flee from human beings is reduced. Their social responses also change: they are tolerant of greater crowding than in nature; they mate earlier and more readily, and they breed for longer, hence may produce more young. Other differences are less obvious. Laboratory rats have smaller brains, livers, kidneys and hearts than those of their wild cousins; and their adrenal glands are also diminished.

All these differences from wild types are shown by the Norways commonly used in laboratories and are often assumed to be entirely genetical; but this is wrong. If docile, domestic Norways are released into some form of freedom, or at least into a large space, their behaviour changes. I experienced this when some of my laboratory rats escaped into the gloomy, junk-filled cellars of a large, ill designed building. When caught and restored to their cages, they were quite vicious. Later I found that Roy Robinson, in his massive survey of rat genetics, had cited early reports of just this effect. More phenomena of this kind appear in later chapters.

We have now turned from tall tales and magic to science. But, of all scientific developments concerning rats, the most important have been not naming, taming or training in mazes, but associating them with infectious disease.

3
ALL FALL DOWN

Ring-a-ring o' roses,
A pocket full of posies.
Atishoo! Atishoo!
We all fall down.
NURSERY RHYME

Humanity today is notorious as a destroyer of species and environments. Yet human settlements have, for more than ten millennia, enabled some species to thrive. Among the mammals are the dogs, cats, cattle, sheep, goats, swine, camels and lamas, all of which have been welcomed. Others, especially rodents, have moved in uninvited; and, with them, have arrived microbes, many of which have become partly adapted to the human body. N.G. Gratz, formerly of the World Health Organization, has published a list, certainly incomplete, of fifty-five infectious diseases derived directly or indirectly from rodents.

MUD FEVER

One is leptospirosis or Weil's disease. If the reader goes down with a baffling fever, headache and heavy sweats, the physician may suspect the condition sometimes called mud fever. And, if the reader has been working in a sewer, or in any other wet place infested with rats, that diagnosis becomes more probable. Leptospirosis is especially recorded among farmers, fishermen, miners and workers in rice fields, dairies and abattoirs. During the war in Vietnam, in the 1960s and 70s, it was the leading cause of acute fever among American soldiers. Gratz holds this worldwide infection to be the most prevalent of rodent-borne diseases.

The organism of leptospirosis is a spirochaet, which is a rather odd kind of bacterium. Spirochaets often live in the kidney tubules of mammals and emerge in their urine. Once excreted, they can survive for some time in mud or dirty water. From there they can enter the human skin, especially if it is cut or abraded.

They have been described as looking like tightly coiled snakes; but they are very small and slender, hence difficult to study under the microscope. They have nonetheless been found in many species of rodents, from the house mouse to the giant rat, *Cricetomys gambianus*—not of Sumatra but of West Africa. One of the highest figures of rodent infection was recently reported in Colorado, USA, where 66 per cent of wild Norways were infected with *Leptospira interrogans*, the organism of Weil's disease.

Among rodents which live in the wild the infection is maintained without human intervention and seems often to be harmless. Human infection has been described as accidental. The problem is a typical example of the tasks of public health: prevention requires getting rid of rats, improved hygiene and other obvious precautions, such as that shown on page 27.

Some measures, however, are less obvious: in AD 2000, in eastern Thailand, during an alarming epidemic of leptospirosis, threatened citizens were encouraged to catch and eat the carriers (probably mole rats or black rats). Methods of trapping the rodents, and recipes for cooking them, were issued by the authorities.

HOW NOT TO HANDLE WILD RATS

I have a personal interest in leptospirosis. In the 1940s, during experiments in a government laboratory, a female Norway was brought in from a rubbish tip at Henley, near London. I lifted her gently out of the trap, with a gloved hand. Though quite docile, she wriggled, crept up my shirt sleeve, scrambled up my back and emerged on the back of my neck. I picked her up more firmly and put her in a cage. (I still remember the expressions of two assistants who were present.)

A week later, I went down with a dangerously high fever, ghastly headaches and profuse sweats; and, after a few days, I began to turn yellow with jaundice—my liver was affected. My doctor was baffled but, at some point, in response to his questions I mentioned the incident with the rat. He thought I was delirious; but he examined my back and found extensive scratches. At that point the penny dropped, and I was launched on a course of enormous doses of penicillin (then quite a novelty). Within a few hours my temperature was down and the headaches had stopped. After three weeks I was back to normal, except that I remained yellow for some time. When I had recovered I was told that Weil's disease had a 30 per cent mortality.

The important feature here is not the rat nor even the author, but the gauntlets. (Photo by Philip Boucas, courtesy WHO)

Since then, when handling wild rats, I have always worn gauntlets, not gloves, and I have insisted on my colleagues doing the same. Today, as I show later, still more rigorous precautions have become necessary.

PLAGUE

Bubonic plague, too, is a story of public health, but also of cultural and economic history and superstition. Like other

27

human relationships with rats, it illustrates the diversity of our social attitudes and practices.

Plague is of several kinds. In its most familiar form a lymph node ('gland') in the groin, armpit or neck enlarges to form a bubo, becomes exruciatingly painful and often suppurates. Fever may follow. Without modern treatment, about 50 per cent of people with these signs die in a few days. This is the Black Death, so called because dark patches appear in the skin. It may develop into another form, septicaemic plague, in which the infection is present in the blood. Fleas can now transmit the infection without the intervention of rats. Survival from the septicaemic condition is rare.

The most terrifying form is pneumonic plague, described in the apparently innocent rhyme at the head of this chapter. (The posy is of flowers or herbs, believed in the Middle Ages to ward off disease and, until recently, ceremonially carried by English judges.) The infection affects the lungs and, without immediate treatment, is virtually a sentence of death: the infected person has about 24 hours to live. During that brief period, coughing and sneezing can transfer the infection to others. An epidemic can therefore occur without carriers other than human beings.

Plague appears in worldwide outbreaks, or pandemics, at long intervals. Accurate information exists only for the last two—the Black Death, which seems to have begun in Central Asia in 1333, and the present one (described later), which began in the interior of China late in the nineteenth century.

In 1333, Honan (modern Henau) and nearby regions suffered a severe drought, followed by famine, floods and locust swarms. One likely accompaniment, due partly to destruction of forests, was irruptions of rodents bearing the plague bacillus. Certainly, in the 1330s and 1340s, plague caused many deaths in a vast area of Asia, including Asia Minor. Living as we do in a time when famines and other disasters are spreading, this history has a warning for us today.

THE BLACK DEATH

When bubonic plague struck Europe in 1347, the idea of transmission by microorganisms was far in the future. (Even the military use of gunpowder was then quite a novelty.) In 1832 a German medical historian, J.F.C. Hecker (1795–1850), published an account which exposes the bafflement of physicians faced by a disease which appears in several forms, each often fatal and all inexplicable. His account, and more recent works by Joannes Nohl and Philip Ziegler, describe, in absorbing detail, the wild explanations offered for the disaster.

Learned persons, especially Christian priests, commonly attributed the illness to the arrow of God (*Psalms* 38, ii, iii), that is, punishment for our sins. This teaching may have provoked the bizarre movement of the flagellants, who emerged in eastern Europe in the fourteenth century and spread over the continent. Its founders are said to have included imposing Hungarian women of great size; but authentic accounts refer only to people who whipped themselves violently and injuriously, in accordance with a prescribed ritual. Their stated objective was to express contrition for their own sins and those of others. At first, it seems, flagellants came from among the poor; but they were later joined by priests and nobles.

Masochism such as this often goes with cruelty. The flagellants took up the custom, already well established, of representing the Jews as demons or agents of Satan who killed Christian children for sacrifice. In the 1340s, on the arrival of plague, the Jews were accused of propagating disease by contaminating wells. According to one story, Jewish conspirators used powdered poisons brought from the East. These legends were encouraged by prominent members of the clergy and led, in Central Europe, Italy and Spain, to massacres in which many Jews were shut in wood buildings and burnt to death. Ziegler comments on the humiliating reflection that the European, overwhelmed by what was probably the greatest natural calamity ever to strike his continent, reacted by seeking to rival the cruelty of nature in the hideousness of his own man-made atrocities.

(Pathological fantasies about Jews did not die with the Middle Ages. In the 1880s, without the excuse of an outbreak of plague, an authoritative Catholic publication printed bizarre stories of the ritual murder of Christian children by Jews. And in 1934, in Hitler's Germany, an official journal published nauseating illustrated libels concerning 'Aryan' children killed by Jews, with rabbis portrayed as sucking the children's blood.)

In fourteenth-century Germany, however, some physicians were influenced by less harmful myths, called 'magic therapeutics' by Johannes Nohl. These were based on the concept of a Universal Soul, which connects all bodies together. Every separate body was said to possess a vital spirit which acted on other bodies. Treatment required the use of a magnetic force, which was held to be present even in human blood and excreta. This led to the use of disgusting (and probably dangerous) mixtures for treating patients. (Today, the notion of a 'Universal Soul' has been replaced by others such as that of Gaia, named from the Greek earth goddess, in which the whole of the biosphere is seen as a single superorganism.)

But, even in the Middle Ages, some explanations dispensed with magic. The most widely held was pollution of the air, which appears, combined with astrology, in Shakespeare's *Timon of Athens*, as

> a planetary plague, when Jove
> Will o'er some high-vic'd City, hang his poison
> In the sick air.

Hecker presents stories of the spread, by a 'pestiforous wind', of poisonous odours, which overpowered people so that they 'fell down suddenly and exspired in dreadful agonies'. And he complains of the 'low condition of science' and of the lack of accurate observers at the time of the Black Death.

In the towns and cities of late mediaeval and early modern Europe, hygiene hardly existed. Hence, as the authorities thrashed around looking for causes, they also blamed stinking fish, bad meat and mouldy corn. F.P. Wilson, in his account of plague in Shakespeare's time and earlier, writes:

> Modern magistrates should know the case of the taverner
> who sold bad wine in 1364, and as punishment was made
> to drink a deep draught of it, and was drenched with the
> remainder.

A rat catcher in Shakespeare's time. (From F.P. Wilson, The Plague in Shakespeare's London, *Clarendon)*

A prominent feature of city streets was an enormous population of dogs. As a result, the dogs (which may have done something to keep down the rats) were sometimes destroyed by municipal order. The poor also resorted to eating the dogs. In the West, rats were hardly suspected as plague carriers, hence dog catchers, but not rat catchers, were well thought of. Yet an ancient Indian text, the *Bhagavat Puran*, had, long before, warned people to leave their houses when a rat fell from the roof, tottered about the floor and died; 'for then be sure that plague is at hand'.

In some communities the total death rate during the Black Death was reliably recorded as above 50 per cent. A quarter of the population of Europe, including England, is believed to have died. It struck more men than women, but made little distinction between classes.

> Sceptre and crown
> Must tumble down
> And in the dust be equal made
> With the poor crooked scythe and spade.

As a result, infected persons were deserted by neighbours, family members and physicians. Some doctors, notably the Frenchman Guy de Chauliac (?1300–1367), a famous surgeon, acknowledged the humiliation felt by physicians when faced with the disease. Hecker admires the 'courageous Guy de Chauliac',

> who vindicated the honour of medicine, by bidding defiance
> to danger; boldly and constantly assisting the affected, and
> disdaining the excesses of his colleagues who ... held that
> the contagion justified flight.

Often doctors could do no more than urge their patients to seek salvation by confessing their sins. Others, however, preferred to emphasise the salutary effect of cheerfulness, instead of preoccupation with sin and death. One famous, or infamous, report concerns the members of a Sanitary Commission of Königsberg in Germany, who, during a tour of inspection, applied this method to themselves. Despite the failure of their treatments, 'with the help of strong Insterburg ale they spent their time in joy and merriment ... drinking, dancing and carousing'. In those circumstances, perhaps it was the best thing to do.

At its worst, the Black Death destroyed whole communities. John Saltmarsh describes the depopulation, economic decline and depression which, during the fourteenth and fifteenth centuries, accompanied the plague. He cites a French historian who wrote of abandoned villages and of farms, fields and vineyards engulfed by the returning tide of the forest. In 1349, in England, some saw the first strike of the Great Pestilence as the final disaster. Saltmarsh quotes a Franciscan Friar, John Clyn:

> Lest things worthy of remembrance should perish with time, and fall away from the memory of those who come after us, I, seeing these many evils, and the whole world lying, as it were, in the wicked one—myself awaiting death among the dead—*inter mortuos mortem expectans*—as I have truly heard and examined, so I have reduced these things to writing; and lest the writing should perish with the writer, and the work fail together with the workman, I leave parchment for continuing the work, if haply any man survive, and any of the race of Adam escape this pestilence and continue the work which I have begun.

After this are the words, *magna karistia*—great dearth; and, in another hand, 'Here it seems that the author died'.

But, in Saltmarsh's words, the race of Adam did not perish. From the Italian city of Florence, struck by plague in 1348, we have a firsthand account by a survivor, the celebrated and controversial poet Giovanni Boccaccio (1313–1375). Many of his works were moralities which urged the value of sacred, rather than profane, love; but, as the historian Judith Serafini-Sauli writes, 'a certain indulgence in earthly delights frequently creeps in'.

Boccaccio was a superb story teller who followed the prescription of cheerfulness in the face of calamity. His most famous work, the *Decameron*, written in 1348–1351 during the worst period of the Black Death, consists of a hundred stories, told by a group of seven young women and three men, members of the nobility, who have fled to a villa in the hills near Florence.

The introduction to the *Decameron* begins with a powerful description of the surrounding disaster, in which Boccaccio emphasises the danger of contagion. Plague, he says, came from the East and, after wreaking unbelievable havoc on the way, had reached the West in spite of all the means that art and human foresight could suggest, such as keeping the city clear of filth and excluding all suspected people. Neither knowledge of medicine nor the power of drugs was of any effect, whether because the disease itself was fatal or because the physicians, whose numbers were increased by quacks, could discover neither cause nor cure.

Infected persons (the poet writes) developed tumours in the groin or under the armpit, some as big as a small apple, others like an egg. Afterwards, purple spots appeared in most parts of the body. People with these signs generally died after three days.

The disease spread daily from the sick to the well. It was not necessary to come near the sick: even touching their clothes or anything they had touched (he believed) was sufficient. A cruel and uncharitable result was that uninfected people avoided the sick and everything that had been near them.

Some felt it best to live temperately but others would deny no passion or appetite they wished to gratify. And the public distress was such that all laws, whether human or divine, were ignored.

Most of the poor were kept at home and sickened daily by thousands. And for lack of service and all else, almost all were inevitably doomed to death. Many breathed their last in open streets; many again ended in their own dwellings and gave first notice of death by the stench of their rotting corpses.

PUBLIC HEALTH

The story is, however, not all of ineffectual action. One outcome of plague, especially in Italy, France and England, was the adoption of measures of public health. In Italy, during an epidemic in Vicenza, belief in the importance of contagion led the municipal authorities to introduce sanitary measures which seem to have reduced the impact of the disease. In England, each parish appointed 'nurse-keepers' to tend the sick. People ill with plague had to be segregated in their houses and warders or watchmen were employed to supervise their isolation. Later a more humane policy was adopted: patients were looked after in specially built pesthouses—precursors of fever hospitals.

A measure of lasting importance and, later, of wide application, was quarantine. Venice, in 1348 the greatest of Italian cities, was one of those most horribly devastated by plague. Measures of hygiene were adopted, such as deep burial

During the last outbreak of plague in Marseilles in 1720, municipal health workers wore elaborate protective clothing and carried flaming torches. (After Johannes Nohl, The Black Death*)*

of corpses; and, from 1374, people returning by sea from the East were isolated for thirty or forty days (hence 'quarantine').

The sequence of plague outbreaks since the fourteenth century is still mysterious. Some writers say that the Black Death soon 'burnt itself out', but this metaphor is unhelpful. After about four years, the infection subsided, for no known cause; yet it returned at intervals, with diminishing intensity, for three or more centuries. The Great Plague of London, in 1665, was its last serious flaring in Britain. The south of France had a final outbreak in Marseilles, in 1720. Both were horrible.

On 16 October 1665, the eminent civil servant and famous diarist Samuel Pepys (1633–1703) was in London:

> Walked to the Tower; but Lord! how empty the streets are and melancholy, so many poor sick people in the streets full of sores; and so many sad stories overheard as I walk— everybody talking of this dead, and that man sick, and so many in this place, and so many in that. And ... in Westminster, there is never a physician, and but one apothecary left, all being dead.

A still more famous narrative of the London calamity, only in part fiction, is given by Daniel Defoe (1660–1731), author of *Robinson Crusoe*. His *A Journal of a Plague Year* is presented as a firsthand description by a Citizen who seems to have been Henry Foe, a London saddler and Defoe's uncle. Among the grim scenes is that of the great pit which was dug in the churchyard in the London parish of Aldgate:

> it was about 40 Foot in length, and about 15 or 16 Foot deep; and at the time I first looked at it, about nine Foot deep; but ... they dug it near 20 Foot deep afterwards.

The narrator describes other such pits.

> Into these Pits they had put 50 or 60 Bodies each, then they made larger Holes, wherein they buried all that the Cart brought in a Week, which ... came to, from 200 to 400 a Week.

So people died by thousands and were buried in unmarked graves; others fled, and left the streets and markets empty and silent.

As in earlier times, the deity was invoked. A physician to Queen Anne attributed the plague to Divine Vengeance poured down on 'a Nation harden'd in Impiety and obstinate in Wickedness'. Another medical authority held that 'Repentance, with a lively Faith in God's Mercies, is the only Remedy'. Louis Landa, in his Introduction to the *Journal*, gives other instances. But he also points out that a near contemporary of Defoe was William Petty (1623–1687), one of the principal founders of medical statistical analysis, who wrote:

> one time with another, a *Plague* happeneth in *London* once in 20 years, or thereabouts; for in the last hundred years . . . there have been five great *Plagues*, viz. *Anno* 1592, 1603, 1625, 1636 and 1665. And . . . the *Plagues of London* do commonly kill one fifth part of the *Inhabitants*.

Modern systems of public health depend on knowledge of the size and structure of the population, the incidence of disease and the causes of death. Eventually, early in sixteenth-century England, while plague still threatened, the Bills of Mortality were established. These foreshadowed modern statistical analyses of illness and death. Plague, it has been suggested, was largely responsible for the foundation of the medical specialty of public health and of the local health authorities which today protect us from infection.

YERSINIA PESTIS

In the nineteenth century came the truly 'epoch making' discovery, in mid century, of the bacterial causes of disease.

39

Vast improvements in public hygiene followed. Yet, as late as 1891, plague was still attributed to soil poisons. In 1894, however, two bacteriologists, a Japanese, S. Kitasato (1852–1931), and a Frenchman, G.A.E. Yersin (1863–1943), in Hong Kong during a plague outbreak, discovered the plague bacillus, *Yersinia pestis* (now known to have several forms). Their findings made possible development of a serum which conferred immunity against plague.

They also led to worldwide researches on the many rodent species which carry the infection. Rodents support battalions of fleas, and human infection usually requires a flea bite. The flea must come from a rodent, often a black rat, especially one that is dying or has died from the infection and is therefore shedding its fleas. By sucking the rat's blood, the insect has acquired the bacilli and these sometimes block its gut. When the insect pierces human skin, it injects a mass of many thousands of bacilli. It also deposits the microbes in its faeces. A person so infected may, however, not notice the bite. Illness develops only after several days.

About 1500 species of flea have been described. Many, perhaps most, can carry the plague bacillus; but, to be dangerous to people, a species of flea must habitually parasitise rats in large numbers, yet be able (if nothing better offers) to pierce the thick human skin. These qualifications are possessed by a famous species, *Xenopsylla cheopis*, which can survive for some weeks without a rat to feed on, and can be carried long distances in goods such as textiles. It may then leap on to people and infect them. Hence bubonic plague can appear in the absence of rats.

Yet, in view of the complications of transmission, it is rather astonishing that the plague bacillus can cause terrifying epidemics. Most of the time, it lurks in rodents of forests and grasslands, where it rarely affects human beings. When it does so, the illness may be called sylvatic plague (Latin, *silva*, a woodland).

The number of 'sylvan' rodent species which can be infected is at least two hundred. Some carry *Y. pestis* seemingly without harm to themselves or their fleas. It is then always present and is said to be endemic or—more properly—enzoötic. (Fleas can also, disconcertingly, survive for months on hibernating field rodents; but this ability seems to have been little studied.) The bacillus moves among the species, breaks out at intervals and produces an epizoötic—the animal counterpart of a human epidemic. One likely consequence of an epizoötic outbreak is spread of the infection. When the bacillus reaches commensal species, especially black rats, it can progress to people.

MODERN PLAGUE

The current pandemic, like the Black Death, seems to have begun when an exceptionally virulent variety of the bacillus arose in Central Asia. It spread with remarkable speed to rodents in other regions, notably India where, in the first years of the twentieth century, deaths from plague were more than a million annually. Then, as usual, it subsided; but it returned in the 1940s; and, as recently as 1994, an outbreak occurred in Maharashtra, in the west of the country, attributable in part to cost cutting by a government trying to reduce expenditure on public health but also to lack of urban hygiene.

Vectors

Central Asia, where the two most recent pandemics evidently began, has many rodent species, including marmots, gerbils and squirrels, all of which often carry the bacillus. In the interior of China, the tarbagan, *Arctomys bobax*, an attractive creature with a valuable skin, has been and is a notorious reservoir which carries the infection to trappers.

In other countries of southern Asia, apparently permanent sources of the plague bacillus include black rats, Norways, bandicoot rats and many others.

In the African continent an important reservoir in the south is a gerbil (*Tatera brantsi*) from which infection moves to black rats and, probably more important, to the small, multimammate rat (*Mastomys natalensis*) which too lives in houses. This exceedingly fecund and widespread commensal may consist of several species. In East and Central Africa, the multimammate and black rats are again widespread, accompanied by other species. Hence Africans, in addition to all their other troubles, must always contend with the threat of plague.

The same applies to South America. There, as usual, the black rat is a leading vector, but it is supported by many other species, including the domestic guinea-pig (*Cavia aperea*) which is a pet as well as (sometimes) a pest. (The genus *Cavia*, alone, has twenty species.)

In the USA, sylvan plague is present in most of the western states and is carried by marmots (*Marmota*), deer mice (*Peromyscus*, of which there are more than 50 species), chipmunks (*Eutamias*) and others. And, as usual, black rats are common.

Throughout much of the world, but especially in northern Europe and the north of North America, Norways are also present and capable of both carrying plague and also transmitting it, via their fleas, to people.

From India it travelled by sea to much of the rest of the world. It came ashore in South African ports around 1899 (when the human inhabitants were distracted by the Boer War between the British and the resident Afrikaners). At first the infection spread slowly but, by the time of the outbreak of the First World War, in 1914, it had reached arid regions populated by many gerbils. Through them, it spread northwards and led to intermittent outbreaks. Zimbabwe had a bad epidemic in 1974.

The Americas too were soon involved. California, especially San Francisco, received *Y. pestis* in about 1900 and organised a massive campaign against its waterside rats. The infection nonetheless spread to independent species and became endemic (enzoötic) or sylvan.

Similarly, in South America, the infection spread from rats in the ports to rodents in the interior. In north-eastern Brazil, where plague is endemic, hundreds of cases are still reported annually. In addition, in the 1980s, plague outbreaks occurred in Ecuador and Peru.

Twentieth-century Europe had two appalling wars but experienced nothing like the Black Death. Plague did, however, briefly visit the east of England in 1910. More serious, in 1945, during the final months of the Second World War, it appeared in Malta. The human population of this group of Mediterranean islands was about 244 000. Of the 80 people diagnosed with plague 22 died.

The special interest of this case was not only the absence of mediaeval panic and paranoia and the use of modern medicine, but also the application, for the first time against plague, of new, scientifically based methods of rat control (described in later chapters). The methods worked. The clearest evidence for this was the disappearance both of plague and of ratborne murine typhus which had long been endemic in Malta. To achieve this, the civil authority employed one operative to every 4000 inhabitants, and the Armed Services many more. Later, it was possible to keep rat numbers low with a small labour force. (For more, see my paper in the *Journal of Hygiene*, 1948.)

NEW VIRUSES

A reader in a rich country, with a modern national health service, may feel that the diseases of rodents are of only remote interest: much of the time, measures of public health prevent them; and bacterial infections, such as leptospirosis and plague, if quickly diagnosed, can be successfully treated with antibiotics.

But other diseases, carried by rodents, are due to viruses, some of which have only recently been identified. J.N. Mills, of the US Center for Infectious Diseases, has described new viral haemorrhagic fevers which, in the 1990s, were being discovered at an increasing rate among both human beings and rodents. Such discoveries are likely to continue.

Like the plague bacillus, the viruses usually move among animal species without touching people. When they do infect us, they again resemble plague in causing great suffering. Fever is likely followed by general worsening of health and by bleeding from vessels in the skin. Later, internal bleeding is probable. The kidneys and other organs may be damaged. Some of these viruses also enter the nervous system and induce severe weakening and seizures. Death is probable after several days.

Viruses do not respond to antibiotics. Worse, a name such as 'hantavirus', which we meet below, does not refer to a single, well defined species of organism. Indeed, strictly speaking, viruses are not organisms, but are usually fragments of RNA (ribonucleic acid) with no metabolism of their own. They can be reproduced only inside the cell of a living animal or plant, or a bacterium. After multiplying and—sometimes— killing the cell, they escape and can be carried in body fluids, water or air to the cells of other hosts. While moving around, they frequently mutate. (The viruses of influenza are notorious for incessantly changing.) In this way they present a continually varying challenge to the human immune system and to the medical services.

Vaccines are available for some of the best identified rodent-borne viruses. Preparing them, however, is itself hazardous. In Marburg, in Germany, in 1967, 25 laboratory workers were infected with an agent, previously unidentified, now called the Marburg virus. Seven died. Today, increasingly stringent precautions are taken by people who work with or near possible vectors—a development which I can whole-heartedly support (see page 27).

One of these infections, Lassa fever, is widespread in West African countries: in the 1990s it annually infected at least 300 000 people and killed 5000. The multimammate rat again seems to be a leading source of human infection, but black rats may help. Other African outbreaks include those due to the Ebola virus which caused much anxiety when it appeared in 1995, fortunately in a form harmless to people, in an experimental monkey colony in Reston, Virginia, USA. In October 2000, when this chapter had already been written, an outbreak in Uganda, in a region ravaged by war and largely occupied by refugees, killed many, including hospital nurses who had not been supplied with protective clothing.

Such infections often arise from the disruption of environments, especially forests, with large populations of rodents. Transmission to people is then not through fleas or other insects but is direct: the rodents disperse the viruses in their faeces and urine (compare leptospirosis, pages 24–7). In 1989, in Venezuela, clearing a forest for cultivation was followed by an outbreak of a strange fever eventually traced to cotton rats (*Sigmodon*). The agent (an arenavirus) was in the dust in the clearings. Even combine harvesters can stir up lethal dust.

South America has at least three other important haemorrhagic fevers, carried by a still incompletely known corps of rodents. In the 1990s, outbreaks occurred in Bolivia, Argentina and Brazil. Of these fevers, one group, called HFRS (haemorrhagic fever with renal syndrome), is due to hantavirus, of which there are several kinds. The viruses attack the lungs as well as the kidneys and—like pneumonic plague—can, in their most lethal forms, kill in a few hours.

Hantavirus first became well known in the 1950s, during the Korean War, when more than 2000 American soldiers were diagnosed with HFRS. The number of Korean soldiers similarly infected seems not to be known but must have been much larger. The spread of infection is attributed to the bombing of forests and the resulting scattering of rodents.

In the 1990s, in Asia, HFRS was reported to infect 200 000 people annually, mainly in China—probably another under-estimate. This worldwide and spreading virus also occurs in

Europe, where it is often carried by Norways. A deadly variety has been found in the south-east of Europe. Another form is responsible for much illness in the north-west, especially among workers who inhale wood dust contaminated by rats.

In 1995, J.P. Webster & D.W. Macdonald did a parasitological survey of eleven farms in England and Wales, with unexpected results. Although Britain has until recently scored well in hygiene, they trapped 510 Norways (a disconcerting figure in itself) and found 4 per cent of the rats to carry hantavirus. (In addition, 11 per cent of the rats carried *Leptospira* and a similar proportion had a species of *Yersinia* which causes an influenza-like illness.)

North America also is widely afflicted and has had a number of outbreaks. The danger is not confined to rural regions. In the wealthy American city of Baltimore, in the 1980s, the large population of Norways was thoroughly studied and proved to be widely infected with a form of hantavirus. About 2000 patients with kidney disease were therefore investigated and many showed evidence of hantavirus infection. Evidently, the virus often enters the kidneys where it is associated with high blood pressure. This finding was made when federal funds and the numbers of people employed on rodent control were being drastically cut and the rat population of Baltimore was conspicuously increasing.

In 1993, yet another infection, the hantavirus pulmonary syndrome, emerged in the south-west of the USA. Like pneumonic plague, it is a rapid and fatal disease of the lungs. In New Mexico, 114 people were infected and 58 died. In the same year, in Arizona, 24 people went down with an at first inexplicable fever. Twelve died. The outbreak is believed to be related to global warming, which has led to prolonged droughts intermittently broken by exceptional rains. The rains allow sudden but brief increases in the food of wild rodents, both plants (nuts) and insects. The rodents and their viruses multiply; but the food supply soon declines and the hungry rodents invade urban areas and infect people. The infection has now appeared also in Latin America.

Rodents, as we now see, make a global, largely underground network of disease vectors. Above ground, modern travel allows carriage of infection over thousands of kilometres in only a few hours. Even the richest and most hygienic countries are therefore at risk from the haemorrhagic fevers. The present narrative will certainly be out of date by the time it is published.

Such warnings of danger are often resented: it is more comfortable to dismiss them as the prejudices of prophets of doom. But we, as human beings, can foresee dangers and, as a result, avoid them. To fail to use this ability is to reduce us to a mindless species, at the mercy of fate and chance.

At the end of his great work, *Voles, Mice and Lemmings*, published during the Second World War, the ecologist Charles Elton (1900–1991) wrote:

> We stand on the near shore of an ocean larger than any that Columbus explored, in which we can at present discern only a few islands rising out of the mist. Let us hope that wise governments will train navigators and equip them to explore more closely the Islands of Vole, Mouse and Lemming.

The need for such research has greatly increased since 1942. Much is being learned. International bodies, especially the

World Health Organization, can now quickly identify strange and frightening infections. Progress is also reported in designing drugs or vaccines which attack the viruses. But, while gigantic resources are still devoted to 'Star Wars' or similar programs, the scale of current research is unlikely to match what is needed.

PART II
QUESTIONS AND ANSWERS

If we wish to understand animals, or to prevent them from infecting us or stealing our food, we need to know them as they really are, not to describe them as though they were either human or possessors of magical powers. When we achieve a true account, we can sometimes also use our findings to suggest what to look for in ourselves.

4

A BATTLE OF WITS?

They are diabolically clever animals.
A.H. BARRETT-HAMILTON & M.A.C. HINTON,
A History of British Mammals

A farmer finds his stored grains stolen or damaged. The thieves prove to be what he calls 'brown rats'. The creatures can be seen every night, running along well marked tracks between burrows and food. At some risk to his fingers, the farmer sets a number of powerful snap traps, garnished with appetising scraps of food, on the runways. The rats disappear: not one is trapped. A neighbour, similarly afflicted, puts down not traps but piles of attractive food laced with a powerful but tasteless and odourless poison. The result is the same: no rats turn up and the bait remains untouched. It seems that the rats immediately identify the ill intentions of their human enemies; so they sheer off.

These are not fairy tales. Uncounted farmers, warehousekeepers and others have had similar experiences. So it is hardly surprising that even zoologists have called rats 'diabolically clever'; or that an authority on the biology of rodents has described human efforts to get rid of rats as a 'veritable battle of wits'.

WHAT DO RATS REALLY DO?

Yet the idea of rats as superintelligent is, on the face of it, absurd. It sounds like a result of the compulsion, which we all have, to describe animals as though they were human.

How then does a modern zoologist test rats' intellectual powers? One thing she does is to watch what they do. I say 'she', less on behalf of feminism than because the earliest reliable and detailed account of rats' responses to strange objects, such as traps, was by Monica Shorten. She was one of a group of zoologists in Oxford University who, in the 1940s, were pioneers in founding wild rat ethology. (By 'ethology' I mean the science of animal behaviour.)

The Oxford group was not the first to publish relevant facts. That honour should perhaps go to Robert Smith, rat catcher to the English Princess Amelia. His work of 218 pages, printed in 1768, was entitled *The Universal Directory for Taking Alive and Destroying Rats, and all Other Four-Footed and Winged Vermin, in a Method Hitherto Unattempted, Calculated for the Use of the Gentleman, the Farmer and the Warrener.*

Smith trapped rats alive in cage traps which, for some days, had not been set but left open, with food in or near them. The rats could therefore first become used to going in and out of the traps. When he eventually closed the traps, many rats were caught. (Smith's method remains useful: I have often used it to trap rats for research.)

One hundred and seventy years later, R.E. Doty made observations rather like those recorded by Smith, but much more extensive and precise. He also carried out large scale poisoning in Hawaiian cane fields which were infested with

Sugar feeds several species of rodents in each of the countries where it is grown. Early scientific findings on feeding by rats were reported in 1938 from the Hawaiian cane fields. (Courtesy R.E. Doty)

three species of *Rattus*, and he did pioneering work on estimating their numbers. His work, however, published in the *Hawaiian Plant Record*, received little attention for some years. He kindly supplied me with the photograph above.

The crucial observations were made in wartime by Dennis Chitty and his Oxford colleagues. They were largely concerned with the responses of wild rats to food, including poisoned bait. In one typical experiment, freeliving Norways were nightly eating about 2 kg of plain food, left on the floor of a barn. On the twelfth day, the food was put more tidily on tin trays and, for 24 hours, the rats almost stopped eating. In this case, the deterrent consisted of harmless, unobtrusive metal objects. Sometimes the effect is even greater: if food is offered in a strange container, it may be refused for several days.

Many similar observations were made, but great care was taken not to draw hasty conclusions. For instance, were the rats 'instinctively' detecting the odours of their human enemies? This popular idea could be ruled out by washing the objects used and handling them with gloves or tongs.

The conclusion was this: wild Norways avoid *any strange object in a familiar place*. The object may be dangerous or harmless; it may even be a pile of nutritious food.

A warning is now needed. It is sometimes convenient to use a decline in food consumption as a measure or index of the avoidance of new objects. But the automatic, mindless recoil, shown by wild Norways and some other species, is an avoidance of any new object, edible or inedible, usually at a distance; and it is the principal explanation of their popular reputation for intelligence. As we see later, this finding was important for devising methods of poisoning. (The apparent likeness to the unthinking conservatism of some human beings, *Homo sapiens*, should be disregarded.)

NEOPHOBIA: CAN CURIOSITY KILL THE RAT?

The avoidance of novelty was first named New Object Reaction; but when I began to work on it I renamed it neophobia. For, despite their apparent simplicity, the early observations raised

Secret Intelligence

The researches in Oxford University were carried out, during and after the Second World War, by members of the Bureau of Animal Population. Until 1939, this group had been trying to sort out the mysterious fluctuations in numbers of voles (*Microtus*) and other rodents, such as the supposedly suicidal Norwegian lemmings (*Lemmus lemmus*). (These may be added to the list of plague carriers met in the previous chapter.) A notable feature of their work was its concentration on *counting* animals. It was said that the creatures they studied did not behave: they only numbered off from the right. The war changed that.

The founder and director of the Bureau, Charles Elton, describes how, when war was impending, he proposed to the British Agricultural Research Council a plan for research on rodent pests. As a result, for eight years Elton and his colleagues worked largely on the biology and management of wild rats; and for five years their findings were treated as secret and, for the most part, not published.

It would be agreeable if I could now introduce a drama of a mole from the German *Sicherheitsdienst*, burrowing below Oxford's dreaming spires, in search of a scientific solution to the rat problem. But, if any such person existed, his reports have not yet been released.

The results of the work in Oxford appeared at last in 1954, in two massive volumes edited by Dennis Chitty. (A third volume, on the house mouse, *Mus domesticus* vel *musculus*, was edited by H.N. Southern.) Half a century later, Elton's Introduction, together with many of the Bureau's methods and findings, remain valuable.

questions. Are the features which provoke avoidance always *objects*? For that matter, just what is a new object?

Perhaps the odour or flavour of an unfamiliar food is a deterrent. 'Flavour neophobia' does in fact probably exist, but it seems to consist only of a momentary hesitation. I have not found clear evidence that a strange odour or flavour can, simply by its strangeness, provoke any prolonged failure to approach a novel object or food. Some odours, however, are disliked: that is, they are deterrent whether strange or familiar.

A brief, hesitant response to new odours is displayed even by laboratory rats, but these highly domesticated creatures are not neophobic. For some time, this created difficulties for experimental psychologists, who were accustomed to regarding laboratory rats as *the* rat. Yet, in this as in other respects, the difference of the domestic varieties from the wild type is very striking. Kept in small cages and offered food in a strange container, domestic rats quickly explore the container and continue to eat as before. Wild Norways, in identical conditions, may remain at the back of the cage and stop eating for days.

NEOPHILIA AND NATURAL SELECTION

For anyone familiar with either domestic rats or wild mammals generally, the neophobia of wild rats may seem rather weird. Left to themselves, mammals often display the opposite behaviour. In 1874 Charles Darwin wrote: 'Animals manifestly enjoy excitement and suffer from ennui, ... and many exhibit

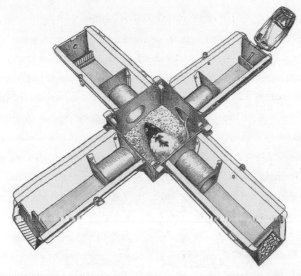

This residentual maze allows experiments lasting many days. An animal has a nest in the central box. Food, water, nest material or other objects can be offered at the ends of the arms. Visits to the arms, and duration of stay, are recorded and analysed by a computer.

Curiosity'. That is, they are not neophobic but neophilic—*attracted* by novelty. And wild rats, despite their neophobia, are, except in special circumstances, quite typical: like other mammals, they too are neophilic.

This apparent contradiction can be resolved in the maze shown above. Put a wild Norway in the nest box, with three of the arms open. It will soon enter each arm, whether it contains food, water or nothing, and it will continue to make regular visits ('patrolling'). After some days, open the fourth arm. This too will be quickly and repeatedly entered.

In all this, the animal is displaying typical exploratory

The Residential Maze

For decades, experimenters on laboratory rats often used mazes to study how rats learn their way about. Their subjects were rarely in a maze for more than a few minutes. When I began work on exploration, I needed an artificial environment in which an animal could live for many days. Once in the system, they could be observed without further handling. (All handling has a disturbing effect, even on domestic rats.) This could, I hoped, give more information on activity and the causes of approach and avoidance. (I also wanted to study the effects of cold on activity; but that is another story.)

Since no suitable device existed, I had to design one. I began with a straight alley, crossed by a light beam which was broken every time the resident went out or back. (Experimental psychologists cheerfully call such a structure an I-maze.) But the findings were difficult to interpret.

While I was struggling, Lee Kavanau, in Los Angeles, developed an enormously complex, automated artificial environment, with many runways, which enabled him to study the movements of deermice (*Peromyscus*). Each mouse learned the system bit by bit, and returned home without error after exploring a new section. Kavanau likened the accurate return to a pianist learning a piece and then playing it backwards.

Eventually, I arrived at the plus-maze in which each arm is crossed by two infrared beams (undetected by the resident animal). The numbers of visits and duration of stay in each arm were automatically recorded and the information analysed in a computer. Once I realised how much could be done with this simple system, I stopped being envious of Kavanau's lavish research grants.

behaviour or neophilia. Now put a strange object in one arm: visits to that arm will stop for hours or days. This is typical neophobia: the avoidance of a strange object *in a familiar place*. As Rick Wallace has shown by experiments with plus mazes, it is avoidance at a distance. Such behaviour protects rats from traps and poison baits (*new objects*), without depriving them of the benefits of exploring new regions.

Evidently, neophobia is a result of natural selection in human settlements littered with traps and poison bait. Or is it? Such assertions often sound plausible; yet we cannot travel back in time to watch evolutionary changes. It is often tempting to say that the evolution of a trait has been observed, even when it can be studied only when performed by existing forms—sometimes, by only one species.

In science we try to make statements which can be tested; and in this case one kind of test is possible. Rats of some species have probably been living with our ancestors at least since a settled agriculture began. This gives them more than 10 000 years for adapting to humanity. If neophobic behaviour is due to human attempts to kill rats, it should be most highly developed in these commensal species—those that depend on people.

The original studies of neophobia were on wild Norways or on black rats, both of which depend almost wholly on human communities. Two other commensals, the soft-furred field rat (*R. meltada*), a widespread pest of crops in India and Sri Lanka, and the bandicoot or Indian mole rat (shown on page 17), are typically neophobic.

Many other species are free living. Australia offers several. My colleague, Phil Cowan, in the Australian National

University, compared the black rat with two noncommensal species. One was the longhaired rat, *R. villosissimus*, which usually lives unobtrusively in the outback. Occasionally, it breaks out into massive and alarming 'plagues' in grazing areas, where it can be trapped for research. The other was the bush rat, *R. fuscipes*. Members of each species were settled comfortably in cages, with plenty of food. After some days, the food containers were changed. The black rats, as predicted, then stopped eating, but the two noncommensal species were unmoved. Later, Gabriele Bammer and others made similar observations on another independent species, the Australian swamp rat, *R. lutreolus*, in plus mazes. So we have real evidence of neophobia as due to natural selection among the rats that enjoy human hospitality.

Should we then classify neophobia as something fixed in the very nature of a Norway—that is, as 'instinctive'? No. An American zoologist, Bob Boice, made an instructive observation on wild Norways infesting a landfill in which everything was, so to speak, a new object. These rats were found to be hardly neophobic. In their unstable environment, they had developed a novel response to novelty: they had habituated to incessant change. Evidently, if most of the objects in a rat's environment are unfamiliar, the tendency to recoil from a strange object is lost.

We can now also see more clearly why rats have been called diabolically clever, and why today we say that they are nothing of the sort. This conclusion was foreseen in 1898 by an iconoclastic American psychologist, E.L. Thorndike (1874–1949). (I owe the quotation to Jeff Galef.)

> Most of the books do not give us a psychology, but rather a eulogy of animals. They have all been about animal intelligence, never about animal stupidity ... [They illustrate] the well-nigh universal tendency in human nature to find the marvelous wherever it can.

We meet more of this 'universal tendency' later. Much more also needs to be said about intelligence and stupidity. But how can these qualities be studied?

5
DO RATS THINK?

Home-keeping youth have ever homely wits.
SHAKESPEARE, *The Two Gentlemen of Verona*

In the first part of this book, and in the previous chapter, rats appear principally as enemies; but, during the twentieth century, experimenters have turned them into collaborators. Many tens of thousands of publications have appeared on what domestic Norways can learn to do in laboratories. In strange, contrived situations, in which the animals at first appear as little more than preparations or mechanisms, their actions soon prove to have features quite unlike those of machines. But, while psychologists were (sometimes reluctantly) facing novel complexities, extremely simplified stories about behaviour were spread outside the laboratories, and mechanical explanations of animal action came to exert a vast, unmeasured social influence.

Hence we now, as in chapter 1, meet rats (with some contributions from dogs) interacting with people in strange and diverse ways.

FINDING THE WAY

Experimenters on rats' abilities often hoped to reveal, in the long run, something about the complexities of human intelligence; but, quite sensibly, they first contrived simple situations.

One device, the maze, seemed to match the habits of freeliving Norways: typically, they live in systems of branching burrows and, above ground, they move on fixed paths next to a wall or other structure. How, it was asked, do they find their way about? More generally, what and how do they learn? The many resulting experiments led a leading American psychologist, E.C. Tolman (1886–1959), to write:

> I believe that everything important in psychology (save perhaps such matters as the building up of a superego, [or] those that involve society and words) can be investigated ... through the experimental and theoretical analysis of the determiners of rat behavior at a choice point in a maze.

Tolman's overstatement need not put us off maze experiments. In a simple case, a hungry rat is put at a starting point and allowed to find its way to a reward of food. At first, it moves around (or explores) without obvious direction; but, after a series of trials, it runs rapidly to the goal without error.

This unsurprising outcome led, after many exertions, to surprising findings. Usually, rats are run in lighted mazes: they then depend largely on what they can see. But they can also find their way in the dark. They are then helped by their

When, as is often does, a rat moves about in contact with a vertical surface, the whiskers, or vibrissae, are important sense organs. They are perhaps also involved in social interactions. (Photo by Philip Boucas, courtesy WHO)

whiskers (vibrissae). Detailed experimenting has shown them to be able also to respond to the gradients of inclined maze floors; and they can use as clues the echoes of their movements when these differ in different parts of the maze.

These are sensory skills. Other complexities concern muscular (motor) abilities. Rats have been trained in a maze and then been presented with the same system full of water. Without further training, they swam through it and so reached the goal by a different set of movements. Others have been trundled through a maze in a small cart. When tested (in a water maze) they did better than controls which had never been taken for a ride. Had they used their eyes to store a map of the maze in their heads? We return to this notion later.

STUDENTS IN MAZES: A DETOUR

Maze experiments have therefore shown rats to be less dumb than they seemed to be in chapter 4. Since mazes are used to test rats' intelligence, perhaps—it was suggested—they would serve also for the human species. As a result, in some researches, humanity has seemed again to be involved in 'a battle of wits' against well matched opponents.

Like white Norways, students often make docile experimental subjects; and so they have been treated as representative of *Homo sapiens* and tested in elaborate mazes. Their reward was escape from the maze. Rats were run in smaller systems of the same pattern. The rats scored better than the students. In further experiments, however, the students improved with practice.

These projects were not merely frivolous. The intellectual differences between species have been much debated. Early Darwinism led to the idea of a scale of intelligence in the animal kingdom with, of course, *Homo sapiens* at the top. As late as 1980, a writer, obsessed with the intelligence quotient, went further and proposed a gradation of types, from *Amoeba* to hypothetical extraterrestrial eggheads, each placed by its IQ.

But the range of abilities in the animal kingdom (let alone outer space) is not measurable on a single scale, such as the intelligence quotient. (Still less are human abilities.) An octopus rivals a rat in its ability to identify visible patterns, but it does not possess the mammalian ability to distinguish objects by their weights: its mode of life does not require it. Some subterranean rodent species are almost blind and cannot learn even visual discriminations. Each species or group of species has its own kinds of adaptability, and is specialised for a particular way of living. *Homo sapiens*, in contrast, is marked by an extreme lack of specialisation. Running students is mazes is not really a fruitful line of research.

PATROLLING AND SAMPLING

A small mammal regularly patrols the whole of its living space (its home range). Often, it seems to be searching for something, perhaps for food, water, shelter or a mate. But exploratory movements continue when all basic needs seem to be satisfied (and if the animal does not go to sleep).

The compulsion to explore can be shown without special equipment. Take a laboratory rat from its cage and put it on a bench. It will move around sniffing and perhaps occasionally rising on its hind legs. It will also poke its nose into any available aperture. This is one of the few kinds of behaviour one can confidently present live in a lecture room. We may call it exploration or neophilia; but this naming, as we see shortly, conceals many questions.

Other forms of such activity, some of which may seem to be performed for fun, can be clearly seen only in experiments. Suppose an animal is confined in its nest, without food, for some hours. It is then released in familiar surroundings. During the next three hours much time is spent on eating and drinking; but each meal is followed by a brief patrol of all parts of the environment, including those which contain no reward.

If the environment offers a number of palatable foods, the rat makes a meal of one. After that, as it moves around, it also

An exploring Norway rears up and sniffs the air.

samples all the alternative foods. Sampling seems to be an aspect of the liking for novelty, or neophilia. Obviously, it is helpful for omnivorous animals, but it has not been much studied.

I did most of those experiments with wild Norways in plus-mazes (page 59). But much of our knowledge of animal exploratory behaviour is owed to psychologists' studies of domestic Norways. Their findings have made a major achievement in the behavioural sciences. Some, perhaps, have significance for human action.

When patrolling, an animal tends to select the parts of its range not recently experienced. This can be seen even in a 'maze' shaped like a **Y** or a **T**, which gives the subject only one

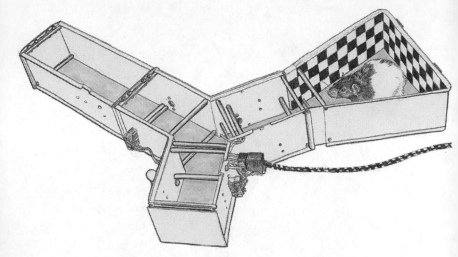

A 'hooded' domestic Norway in a Y-maze has been offered the choice of turning left or right. Its movements are automatically recorded. The design shown can provide different incentives (rewards) in the two arms.

choice. A tame Norway is put in the stem and, if it moves at all (which it usually does), it can move only left or right. If, after it has made a choice, it is put back in the stem, it nearly always walks into the other arm. Even in such cramped quarters, the animal prefers the less recently experienced condition.

If both arms are empty, this spontaneous alternation continues. If, however, one arm leads to a dead end but the other to a spacious, well furnished area, the latter is preferred. Access to a complex environment is rewarding. So is the presence of any novelty, including both unfamiliar food pellets and also strange objects. Many experiments have shown this liking for variety. It is as if rats are easily bored and tire of what they experience every day.

Domestic Norways seem then to reveal two kinds of neophilia: while patrolling, they go for the less familiar; in preferring variety they at first respond to complete novelty.

Mazes, as we see later, have been used to study more baffling behaviour. But, despite Tolman's provocative remark quoted above, psychologists have not always spent their time 'in wand'ring mazes lost'. During much of the twentieth century, research on intelligence and learning was dominated by two other methods and by the two men who designed them. I have written about them in *The Science of Life*, but more always remains to be said. This part of the story is as much about psychologists as about rats.

MECHANISTIC PSYCHOLOGY

The earlier figure was the outstanding (and domineering) Russian physiologist, I.P. Pavlov (1849–1936). He was a principal founder of the science of digestion; but his most famous researches, begun in his fifties, were designed to find out how the brain produces intelligent behaviour.

The conditional reflexes (CR) he measured were attributed to new 'connections' in the brain—an electrical metaphor. His objective was magnificently bold; for his findings were to explain not only the conduct of dogs under restraint but also the most profound achievements of the human intellect. His animals, however, were catheterised and strapped to a table so that their reflexes, especially salivation, could be accurately measured (look at the picture). His method therefore allowed him to record little activity and certainly no intelligent behaviour.

A dog in a sound insulated room, behind one-way glass, harnessed for research on conditional reflexes. (Courtesy Jackson Memorial Laboratory)

The other mechanistic extreme we owe to the most prominent of twentieth-century American psychologists, the benign but obsessed B.F. Skinner (1904–1990). Skinner and his many successors followed Pavlov's example in largely disregarding the behaving animal; but they often went further, for they also insisted on ignoring the nervous system.

Skinner's device is his Box (page 75). The animal to be studied (called S, for subject) is shut in and, usually, has access only to a lever, which can be depressed, and a small bowl. Pressing the lever may or may not release a pellet of food or a drop of water; less often, it may switch on (or not switch on) a source of light or heat. The floor is a grid which can

Pavlov's CRs: the Truth

Summaries of Pavlov's findings often present only an animal salivating in response to food in the mouth and later, after training, performing *the same response* on hearing a sound. Salivation in response to the sound is the conditional reflex (CR). The CR is seen as a simple effect of association with the sound (that is, with the 'conditional stimulus'). The subjects appear as little more than mechanisms; and it is possible to make an electrical gadget which behaves just like that.

But such accounts are highly misleading. They first omit the long training needed before dogs will accept confinement, and their unmechanical differences in responsiveness ('personality').

Moreover, the new, conditional response is *not* the same as the initial response. Many changes occur during training. At first, the dog reacts to the sound by pricking its ears and turning its head (the orienting reflex); as well, its heart and respiratory rates rise. Salivation occurs only when food is in its mouth and is accompanied by chewing and swallowing. As training continues, some dogs paw the ground and lick their lips. Later, they calm down: they no longer perform the orienting reflex and their heart rates and breathing remain steady. Salivation by a trained dog is therefore not accompanied by the responses observed at first. Moreover, it is now anticipatory: it occurs *before* the food and without chewing or swallowing.

Notable findings concern ways in which the CR can be lost. Experimental failure can be due to distraction. An enthusiastic assistant, after exacting labour with a helpful dog, establishes a consistent CR. He hastens to report to Pavlov, who comes to see for himself. The CR then fails. The trouble, of course, is the presence of the prof. The animal has been responding not merely to the buzzer but to the total situation. For consistent results, therefore, the conditions have to be immaculately uniform.

Still more important is extinction. Train a dog to salivate at the sound of a bell. Then give it a series of trials, in which on one day the bell sounds but no food appears. The salivation fades. (No use salivating if no food.) Next day, more trials with no food, but the salivation reappears. Disappearance of the CR in this case is evidently a learning *not* to respond; and *this* habit fades more quickly than the original CR.

The most notorious failure of the CR is the 'experimental neurosis', which seems to be an outcome of the excitement (tension, stress?) that goes with training. (An important feature of the situation is that the animal is strapped to the table and cannot get away.) The animal is trained to respond to a pure tone and is also trained *not* to respond to another tone. The tones are then brought progressively closer, so that the animal has to make an increasingly narrow discrimination. Finally, the task becomes too difficult. The dog becomes incessantly restless *or* catatonic; and, instead of running to the experimental table, resists the experimenter. This is not the reaction of a mechanism.

Many simple responses can, however, be studied by Pavlov's method. (A favourite example in the human repertoire is the eye blink, which protects the eye from damage. A human subject can learn to blink at a slight, harmless sound, when it has previously been coupled with something that threatens the eyes.)

We can therefore see CRs as components of more complex activities. As a mammal moves about or feeds, it is likely to meet situations in which survival is aided by acquiring a CR. But, as we see below, it has other abilities which fall quite outside Pavlov's system.

In a Skinner Box, a domestic Norway presses the lever. Will a pellet of food fall into the bowl? Or the floor grid administer a slight shock? Or will nothing happen? (Photo by Philip Boucas, courtesy WHO)

provide a mild electric shock. A modern experimenter (or E) depends on machinery to provoke action by S and to record S's bar pressing. A computer is programmed to deliver food or shock (or other reward or punishment), often or seldom, according to E's requirements. The program is called a *schedule of reinforcement*.

In a simple experiment, a hungry rat is put in the box. It at once explores, but this activity is not recorded. During these movements, it incidentally depresses the lever (this *is* recorded): there is a click but nothing else happens. The animal continues to move about and, in passing, presses the lever again; and again. On (say) the fifth occasion, with a slightly different sound a small, sugary pellet drops into the bowl and is eaten. The animal has been rewarded.

If a hungry rat receives a donation of food once in every five bar pressings, it rapidly presses the bar five times, accepts the reward, again presses five times and so on. But if food arrives only once in a hundred pressings it is likely to stop pressing for long periods. An alternative schedule is delivery of food *on average* once in (say) 100 pressings, but at random intervals; the animal then persists in pressing and makes few long pauses. In such a situation, a human being would say that one cannot tell when the payoff will occur. In the accounts of such experiments, achieving a reward is described as a result of trial and error (and, eventually, success).

More surprising, an animal can be trained to *delay* its response. Suppose pressing the bar releases a pellet, but no further reward occurs *unless* the subject waits for ten seconds before pressing again. Eventually, it may learn restraint. (I have not done research of this kind; but, while teaching undergraduates, I succeeded in training an S to wait. My Skinner Box had a peephole, and I watched what S did. While waiting, it ran round in circles.)

Some still less obvious findings come from the effects of shock ('punishment'). It is of course possible to train an animal to avoid the lever by giving it a slight shock when it presses it. But a more severe shock disrupts behaviour and the training is then less effective. Moreover, if a mild shock is given *with* the reward, S may be positively encouraged to perform the rewarded act. As many teachers and criminologists know, the effects of threat or of 'chastisement' on human beings are often still more difficult to predict.

The reader may be impressed by the ingenuity of Skinner's method which allows many programs of rewards and punishments and yields numerical findings of great precision; but one may also wonder at a procedure which does not record what, apart from pressing a bar, the animal actually does.

DO RATS GET BORED?

A problem for behavioural scientists, therefore, as for other researchers, is to design not only ingenious devices but also—much more difficult—to ask fruitful questions. As we know, rats, including wild Norways, welcome novelty: they are neophilic. If they are released into a new environment, even if they have fasted for hours they move around and, at first, disregard any food they find. Priority is given to exploring.

Many experiments have been done to find out just what kinds of variety rats will work for. By an anomaly, the Skinner box has helped experimenters to answer such questions. It is an anomaly because their experiments have shown that habits are often not developed in the way assumed by Skinner: that is, solely by responding promptly to a reward (such as food) or punishment (such as shock).

In a Skinner box, rats will energetically press the lever to achieve no more than a slight noise. More interesting, they will also learn to turn a light on. This is not because they want more light, for they will also work to turn it *off*. Here evidently is again a liking for change. But, after some days, the behaviour alters and the animals are likely to work for a dim light rather than a bright one. Perhaps they have become bored with the changes in lighting and are now voting for a preferred condition.

Most of the relevant experiments, however, indicate a preference for stimulation, rather than peace and quiet. This liking is not a prerogative of Norways. The news of neophilia has increasingly spread among people who have the care of other mammals. What looks like boredom, as Miranda Stevenson has described, is a likely result of captivity for any mammal. She writes of the 'appallingly little scientific research on the effects of different environmental stimuli and changes to the environment on the behaviour of captive wild animals'. But, now that zoos are increasingly concerned with protecting and breeding endangered species, this is changing.

PLAYING TO LEARN

In biology, it is usual to ask of any trait, how does it contribute to the possessor's survival or breeding? For exploration, the answer often seems obvious: it enables an animal to find a mate, or new sources of food, water and nesting material and new shelter. But exploring has other functions, not obvious, which have been revealed by domestic Norways in mazes.

Suppose you put a rat in an unfamiliar maze with no goal: it will move around but nothing else can be seen to happen. Repeat the exercise: still nothing. After several such excursions, make it hungry and run it with a reward (a pellet of food).

It will learn to run straight to the goal more quickly than similar rats (controls) with no previous experience of the maze. Evidently, during those apparently futile wanderings, the rat has stored potentially useful information. In ordinary speech, it remembers its way about, just as we often do in similar circumstances.

Such latent or exploratory learning is more tricky to study rigorously than I have shown, and at first it disconcerted specialists. When they had asked what impels a rat or a human being to learn something, their agreeably simple reply had often been that of B.F. Skinner and was in terms of payoff: learning was held always to depend on obvious reward or on avoiding punishment. Many experimenters (and perhaps some teachers) have taken this presumption for granted. But, as we now see, it is wrong.

Especially in early life, exploring does more than allow latent learning. A familiar feature of youthful behaviour is its restlessness. This includes 'play'. I put that everyday word in quote marks because, although we can usually identify the

Contradicting William

Preoccupation with immediate reward and punishment is an example of the limitations of the principle of that celebrated theologian, William of Ockham (1300–1347?): explanations, he seemed to say, should always be as simple as possible. Although Aristotle and others had already said much the same, it was 'Ockham's Razor' that finally caught on. Yet this is not a scientific proposition: it is an instruction or a dogma. And, if it were wholeheartedly adopted, much science would disappear.

activities we call play, it is often difficult to describe, let alone explain them.

Nonetheless, they are important. An outstanding and original Canadian psychologist, Donald Hebb (1904–1985), once took some young laboratory Norways home for his daughters, Jane (seven) and Ellen (five), to play with—and to play with his daughters. Later he took the rats back and tested their ability to learn the way to food in a maze. The petted rats did better than their cousins which had been kept in the usual boring cages.

This was hardly rigorous research but, later, many properly designed experiments were performed in Hebb's laboratory. I mention above the preference of domestic rats for an environment which contains a variety of objects. Young rats are energetically exploratory and playful (and, in a complex environment, they often have to be retrieved by mother). Hebb's domestic researches had suggested that the resulting stimulation influences the *development* of ability.

In some careful experiments young rats were therefore given much experience of moving about a spacious, heavily furnished area. This, as expected, had a favourable effect on their later ability to learn to run a maze, especially if the maze was in the dark.

Some kinds of early experience have special effects. Rats have been reared from birth in cages with black circles and triangles on white walls. Others, the controls, were allowed only a featureless background. All were later required to distinguish circles from triangles in the apparatus shown opposite. Those exposed to modern decor in early life did better. (It does not follow, however, that parents of young children

If the animal jumps at one pattern, the door opens and reveals a reward of food; at the other, it falls into a net. The ability to learn quickly to distinguish such patterns depends on experiencing similar visual stimuli in early life.

should insist on *modern* decor in their nurseries. More on human implications later.)

Early learning to learn, or deutero-learning, seems to be quite different from acquiring a habit, such as running from a starting point to a goal. Hebb describes it as inefficient and slow; yet, later in life, it enables an animal quickly to adapt its behaviour to new circumstances. In ordinary terms, it enhances the development of intelligence.

THE QUESTION OF THINKING

In the previous chapter, I dismiss the idea that rats are 'diabolically clever'. Yet, as we now see, even domestic rats display a notable range of abilities. Hebb writes:

> In mammals even as low as the rat it has turned out to be
> impossible to describe behavior as an interaction directly

between sensory and motor processes. Something like *thinking*, that is, intervenes.

Hebb is criticising a simple assumption, which was once widely made: that, when an animal has learned the way through a maze, it has acquired a fixed set of movements: *run 9 cm, turn left, run 11 cm,* and so on. But, as we know (page 65), rats can learn a maze by performing one set of movements, and then 'run' it correctly with different movements. Their behaviour resembles that of the reader who knows the way about a countryside or a city and can therefore reach a goal by novel means. One speaks of having a map in one's head. So, evidently, have rats.

Hebb's assertion has been reinforced in other ways. Suppose that, on several occasions, an animal simultaneously *hears* a tone and *sees* a light go on. A human being might say that the noise and the light seem to be linked; but the animal says nothing and shows no sign of having learned anything. Next, the tone is sounded alone on several occasions and each time is followed by a slight shock: the shock makes the animal freeze (or display fear). After such training, the tone is sounded without shock, and it still induces freezing, as in a classical conditional reflex.

Now the *light* is switched on alone; and, *although it has never been followed by shock*, it too induces freezing. Evidently, without reward or any other outcome, an association had been silently established between the light and the sound, when they were presented at the same time. This enables the animal later to put two and two together. A human being, in a similar situation, might say that the light 'meant' shock.

Hence psychologists write of the cognitive element in such learning.

An expression, 'cognitive maps', had already been used in famous studies of direction finding. Rats first learn to run to a goal by a roundabout route. They are then allowed to choose from several possible alternatives and, without map or compass, they usually take the short cut which leads directly to the goal; yet they have not taken this route before. Norways evidently have a highly developed capacity for navigating by visible cues, sometimes quite distant. In another kind of experiment, a rat learns the way to a goal and is then set down *at a new starting point*: it is likely to run straight to the goal, again by a route not previously used. Moreover, it can learn new routes, if carefully trained, day after day. (For more, consult Etienne Save.)

In chapter 4 I show how describing rats as if they were human can lead to absurd error. Yet the expression, cognitive map, shows how the behaviour of even so 'lowly' a mammal as a rat is often conveniently described in words derived from what we know about ourselves. Certainly rats often *seem* to think. Understanding what they get up to therefore entails a struggle to match methods, theories and words to the real complexities of animal behaviour.

ARE PEOPLE EVER LIKE RATS?

Another, more prominent struggle concerns our search for clues from animals during our attempts to understand ourselves. We have records, covering at least two-and-a-half millennia, of explaining or expounding human action by resort to animal stories. Dogs strapped to tables or rats in boxes provide special cases of this zoomorphism. The behaviourist systems of Pavlov and Skinner are special because they seem to carry the imprimatur of Science. Though remote from the world of nature, they have sometimes appeared to offer answers to difficult questions about humanity.

Pavlov's researches gave behavioural science much detailed information and a new vocabulary. They did not, however, (as Pavlov had hoped) tell us how the brain works. Yet, despite their complexities, the expression 'conditioned reflex' (a mistranslation) became part of colloquial speech. It signifies a procedure in which, as a result of experience, a stimulus is followed promptly and 'automatically' by a response such as an eye blink. (Sometimes it refers to an unthinking, and perhaps stupid, response to a problem.) Thinking (or cognition) did not enter the picture; nor did feeling. Yet Pavlov wrote, of his own species: 'habits based on training, education and discipline of any sort are nothing but a long chain of conditioned reflexes'.

We should not pillory Pavlov for overestimating his findings. All dedicated researchers are liable to do that. More open to criticism are the leading thinkers who uncritically adopted the CR. The eminent philosopher and polymath Bertrand Russell described CRs as 'characteristic of human intelligence'; and he stated that 'our intellectual life, even in its highest flights, is based upon this principle'. He even held learning to speak to be acquiring a series of CRs. Yet nearly every sentence we utter or write is a new one. Such incessant improvisation, which is part of our everyday life, cannot be fitted into Pavlov's system.

Some people, perhaps, are happy to be likened to attractive

animals such as dogs. Rats are less agreeable. Yet, like Pavlov's CRs, Skinner's ideas, based largely on tame rats in boxes, have had an impressive social impact. One of his followers, R.E. Ulrich, has written:

> Virtually every institution in the United States has been
> touched by behavior modification, if only in feeling a need
> to erect defenses against it.

'Behavior modification' here stands for applying, in schools, factories, clinics and prisons, a system in which the human intellect and emotions are reduced to the consequences of immediate reward and punishment. Skinner, late in life, exposed his limited vision when he wrote:

> Early man ... did not look at pictures or listen to music ...
> When he had nothing to do, if we may judge from related
> species, early man simply slept or did nothing.

These statements are without foundation. Once their primary needs have been satisfied, mammals, as we know, often display a restless curiosity; and human beings do so still more: the ill effects of depriving people of stimulation, as in solitary confinement, have been closely studied for decades. Moreover, the efforts of 'early man' in creating works of art on stone or bark are famous. None of this fits Skinner's presumptions.

THE TREE OF KNOWLEDGE

But perhaps the discovery of cognition (or thinking) by rats can help us to understand ourselves? For we do find, in human action, similarities to what we know of stimulation, exploring and curiosity during the early development of mammals. These similarities, in themselves, *prove* nothing about 'human nature', but they encourage us to study corresponding phenomena among our children.

Harmless experiments have been done with infants aged only two days. The American psychologist Jerome Kagan showed them lights that moved or flickered. He also offered them patterns with contrasts of black and white. All these attracted attention. Hence even at this age changing stimuli are already welcomed. Other experimenters rigged pillows so that babies could cause movement in a mobile if they wriggled about. They soon learned to wriggle in the correct way and obviously enjoyed the result. Babies also give special attention to slight differences in familiar objects, including faces. In this way, it seems, they develop the remarkable ability, which we all have and take for granted, to identify an object from any angle and over a range of distances. Infants have also been tested with sounds. A scale is played on the piano several times; then the same notes are played but in a changed order. The infants smile and gurgle.

Should we then energetically stimulate our infants and young children; and, later, should we strenuously encourage curiosity? In European thought, curiosity about the world has had a mixed press. In *Genesis*, Eve is forcefully warned off the fruit of the tree of knowledge; but she ate it nevertheless and gave some to her husband.

And the eyes of them both were opened.

The millennia since that was written have seen an unending debate between supporters of independent investigation and unfettered argument and those who advocate a society based on the inculcation of dogma.

Today we often encourage curiosity and individual thought, especially in children. This policy can be supported by evidence. Although children cannot be kept like rats, in cramped and featureless cages, we can study the various ways in which they are reared in different communities. An American anthropologist, Berry Brazelton, has observed in great detail the young of an isolated montane people in South Mexico. In their cold climate, the infants and young children are kept wrapped up and given little stimulation or scope for investigating their world. When studied in early childhood, their scores in physical and intellectual development are lower than those of American children. He also describes the adults as imitative and suspicious but able to work for long periods without relief. These are interesting findings; but, as Brazelton himself points out, they allow no firm conclusions. The children are, on modern standards, not well fed. They are also brought up in traditions quite different from those most readers would consider normal. They do not make a satisfactory experiment.

The same applies to children in another harsh environment, the highlands of Guatemala, studied by Jerome Kagan. During their first year they are restricted to their parents' hut. They are also, like Brazelton's subjects, not well nourished. In conduct they are withdrawn and unsociable; and, on most behavioural tests, they are retarded. During their second year they are allowed out and become more mobile. By their eighth year they are described as very active; and by the eleventh year their development differs little from that of American children.

Kagan comments on the 'tape-recorder' theory of human development, according to which the effects of early conditions cannot be erased. He rejects this notion; and his rejection is supported by observations made on orphans from Korea who were adopted by American families. These children, victims of war, who had been hideously deprived in early life, benefited conspicuously when given exceptional stimulation by their foster parents. Hence, despite the impossibility of doing rigorous experiments, all the evidence points to children's need for plenty of freedom to experience agreeable stimulation.

THOUGHT ABOUT THOUGHT

So studies of other species, even domestic Norways, can suggest useful ideas about the human species. But what of the question that heads this chapter? Can we reciprocate, and use our knowledge of ourselves to explain what animals do?

In ordinary speech, animals—especially the mammals that live with us—are allowed wishes, beliefs and understanding much like our own. This practice, as the ethologist Colin Beer has described, is called folk psychology; and the previous chapter shows how it can sometimes lead to error. The behaviourism of Pavlov, Skinner and many others was an attempt to replace colloquial speech with reliable laws of behaviour. It was also at first a legitimate search for simplicity, but it was defeated by the complexities of actual behaviour— indeed, as we have seen, by that of the domestic Norway, let alone the rest.

In response, later generations of brave behavioural scientists have joined in rejecting folk psychology but have developed a 'cognitive ethology', designed to give a scientific account of the *minds* of animals. In an earlier period, many workers in this field would have met such a notion with scorn or derision. It was so greeted by Skinner as late as 1987. He quotes a famous verse (author uncertain):

> The Centipede was happy quite,
> Until the Toad in fun
> Said 'Pray which leg goes after which?'
> And worked her mind to such a pitch,
> She lay distracted in a ditch
> Considering how to run.

And he suggests that the sorry centipede was a victim of cognitive ethology: she would have done better with no mind, but only a set of well ordered reflexes.

Cognitive ethology recognises that the information received by an animal through its senses can be stored and altered in the brain and then recovered and used. (How the brain does this is not known.) But, after that, it meets difficulties. If animals may be said to have minds, these are presumably not like ours. The philosopher Ludwig Wittgenstein (1889–1951) famously wrote, 'If a lion could talk, we could not understand him'.

The difficulties faced by theorists of learning have been put rather differently by Sam Revusky. The passage below— a notable combination of arrogance with modesty—is on 'the traditional study of animal learning'. It was published in 1977 but is still valid.

... people have been working on [learning] for over 75 years and it is still almost in a state of chaos. Just recently, an eminent learning theorist (me) was capable of writing that 'the factors which can supply coherence to learning are ... ephemeral ... and there is remarkably little agreement about the nature of the learning process'. No other basic biological process has had so much attention devoted to it, as evidenced by countless volumes of learning-related journals, with so little in the way of generally accepted basic concepts.

Correspondingly, we have no generally accepted vocabulary for describing animal learning or intelligence. Hence my answer to the question in the chapter title is equivocal: they sometimes *seem* to think. That is why, in the present book, when I am describing behaviour I often use 'mentalistic' expressions: for instance, I may say that an animal evidently 'prefers' or 'likes' something or seems to 'dislike' or 'fear' something else. This way of speaking becomes still more convenient in the chapters that follow.

6

ARE RATS GLUTTONS?

Our natures do pursue
Like rats that ravin down their proper bane,
A thirsty evil; and when we drink we die.
SHAKESPEARE, *Measure for Measure*

If rats do think, it must often be about food. Like other rodents, they are especially equipped to cope with hard things, such as wheat grains: they can even gnaw through lead pipes. Yet Norways and others, with access to many foods, may seem quite indiscriminate.

They can be predators. Some feed largely on snails. Many wild Norways live in the banks of lakes and streams, especially if the water is full of fish. In one such place, food was regularly thrown in the water and attracted the fish. It also attracted the rats: they assembled punctually at feeding times and ate both the food and the fish. (Norways swim well.) At a roost of starlings (*Sturnus vulgaris*) in eastern England, Norways regularly moved into nearby banks when the birds arrived: they were active at night and attacked any bird that fell to the ground.

When I watched colonies of wild Norways in partial freedom, they were fed largely on wheat grains; but, offered

91

A wild Norway chewing a wheat grain. Handling by rats is used in eating and nest building but has been little studied. (Drawn by Gabriel Donald from a photograph)

cabbage leaves, they quickly ate them; and raw meat was also avidly consumed. The largest wild Norways I have known had been feeding on partly buried carcasses in the grounds of a glue factory. Some had reached 700 grams—hardly, in the conventional phrase, 'as big as a cat', but impressive nonetheless. (The factory owners got into trouble with the local health authority.)

SELECTING FROM A MENU

Yet, in experiments, rats have also shown great powers of discrimination. Their choice of food may even depend on what their neighbours eat.

For an omnivore, sampling is important. A rat offered several foods makes a meal of one but, afterwards, briefly tastes the others.

For rats, as for us, a first necessity is to get enough. In the short term, adult Norways adjust their energy intake to maintain a steady weight. But, like many people, they continue to expand slowly throughout most of their lives: over weeks and months they maintain a steady growth. In a Skinner Box (page 75) they can be regularly rewarded with food; they then press the bar at rates which match the target weight. Similarly, if they are given food with much added indigestible fibre, they increase the volume consumed to keep up their energy intake. But, if digestible oils or fats are added, consumption goes down to compensate for the extra energy.

In stable conditions, rats eat in darkness, a few grams at a time, at regular intervals. The amount eaten at each meal is adapted to need. They can, however, alter this pattern. I.R. Inglis and his colleagues have described variations in 'personality' (they do not use this term) shown by the feeding patterns of wild Norways. Individuality, as I have seen myself, extends to small details. One rat, eating flour, may bury its nose in the food, while another sits on the edge of the trough dexterously scooping the food into its mouth with one paw. Within a colony of wild Norways, individual meal times may be influenced by dominance relationships: as Manuel Berdoy has shown, the oldest or heaviest rats have priority at food sources.

Rats' mealtimes can also be drastically altered by training. When I was in Calcutta in 1970, mole rats could be seen, in brilliant sunlight, emerging from their burrows in a much frequented park where they were fed by passers by with grain sold by street vendors.

Human beings, like rats, are omnivorous. Hence we are regularly and rightly told of the importance of a balanced diet. One of the foremost achievements of twentieth-century science has been the discovery of how to design one. Rats, too, given access to a variety of foods, often choose 'wisely'. How do they do it?

When experiments on food selection began, eating a balanced diet was vaguely attributed to instinct. Some early workers, dissatisfied with this notion, not only provided a choice of foods but also watched what the rats did. L.J. Harris and his colleagues, in Cambridge (England), made domestic Norways deficient in vitamin B1 (thiamin) and then offered them two differently flavoured mixtures, one with the vitamin. The deficient rats ate much more of the enriched food. Once they were familiar with the situation, they made regular brief, sampling visits to each mixture at every meal. Control rats, not deficient, ate each food indiscriminately, unless they too became deficient. Similar findings have been made on rats short of two other vitamins of the B group, riboflavin and pyridoxine.

The importance of sampling foods is obvious when rats are offered a cafeteria with many options. (This resembles conditions in freedom. The rat chow used in laboratories is a complete diet but is quite unnatural.) Deficient animals tend to choose unfamiliar foods, and so increase their chances of finding the best one. They may also show an aversion to an unsuitable mixture by spilling instead of eating it. Another aspect of aversion, not well studied, is covering the rejected food with material such as earth.

When a human being eats a needed substance, one result— usually after a delay—may be a feeling of wellbeing and the statement, 'I feel better'. Evidently, something similar happens to deficient rats when they take in a vitamin. Like us, they learn from experience of different foods and return to those with a favourable effect. Deprivation can drastically alter preference. A healthy rat refuses salty food or water, but consumes them readily if it is salt-deficient. Similarly, people who sweat heavily lose much salt and develop a 'salt hunger': miners in deep, hot mines drink salt beer (horrid to most of us) and eat heavily salted foods.

It may seem anomalous that rats should indulge in 'squirreling'; but their food habits also include hoarding. At harvest in the south of India, mole rats drag rice shoots into their burrows in such quantities that the men who eat the rats (page 11) also collect and eat the hoarded grains. In experiments by Dwain Parrack in West Bengal, the grain stored was four times the weight eaten by mole rats in the same period. This is not merely a curiosity, for it reminds us of the problems of calculating losses due to rats. Even if, improbably, the numbers of rats in a region were known, losses could not be accurately worked out by multiplying this number by the estimated amount of food each rat consumes.

Rats as Gourmets

Rodents do not always select foods as if they were biochemists. The flavour or odour of some harmless substances puts them off. Wild Norways I have watched were repelled by butyric acid or aniseed oil mixed with their wheat grains. (Some domestic Norways readily accept both.) When they had no alternative to one of these mixtures, they briskly and dexterously picked over the food: evidently, some grains were less contaminated than others, and these were selected.

Texture too is a factor. My wild rats preferred finely divided wheatmeal to the whole grains from which the wheatmeal was made; and edible oil added to grains increased acceptance.

Are these preferences 'innate' (or 'instinctive')? That is, are they fixed in development, regardless of experience? To answer such questions is often difficult. Consider sweetness. Rats, like human beings, are inclined to favour sweet mixtures—not only those flavoured with sugar but also those with saccharin, which has no nutritional value. Perhaps this preference develops regardless of experience. But perhaps not. Milk is slightly sweet; is the liking for sweetness acquired in infancy, at the mother's 'breast'?

Much therefore remains to be learned. Of one thing, however, we may be sure: rodents do not develop food tabus. No murine tribes exist which reject certain foods because they are forbidden by law or morality.

CONSUMING PASSIONS

Philosophers sometimes debate whether we have free will. Whatever they say, human beings often seem to express strong wills when they select foods. When we do this, to what extent are we influenced in the same way as rats?

Certainly, we regulate our energy intake and we begin early. Infants born much below the usual birth weight tend to take exceptional amounts of milk and so grow toward a more usual body weight. Large babies do the opposite. (These statements apply especially to babies fed at the breast, for a mother's milk production is adjusted to demand.)

When we are adults, 'dietary self selection' is not precise from day to day. The amount eaten by some people has a seven-day rhythm: body weight declines slightly during the working week and is made up at the weekend—a cultural influence from an early Middle Eastern civilisation in which rest on the seventh day was mandatory.

Although, therefore, we control energy intake, we surely (the reader may demur) do not select biochemically correct food. Consider all the religious tabus and other arbitrary customs reported by anthropologists. Nonetheless, we do possess some ability to select a good diet, without the help of biochemists.

Children with no strongly developed food habits are perhaps better at food selection than adults. In famous experiments by C.M. Davis, young children were offered, over a long period, a free choice from about twenty foods at each meal. Milk, meat, cereals and fruit or vegetables were always supplied; other items varied. Often, the results were unnerving: for some days, a small child might eat nothing but one food. But these binges ended spontaneously. The choices of some other children were described as 'a dietician's nightmare—for example, a breakfast of a pint of orange juice and liver; a supper of several eggs, bananas and milk'. Yet, in the long run, a balanced diet was achieved: all the children were healthy and grew well. These children were, of course, not offered foods made largely of white flour and sugar and without fibre; nor were they given sugary drinks.

A nutritionist, D.A. Booth, remarks that adults too have some ability for 'automatic tuning' of their diet. But this cannot be relied on. The reader almost certainly has ready access to skilfully advertised, highly palatable foods and beverages which have no place in a balanced diet. To find the way through these hazards, use has to be made, not of 'instinct' or of 'conditioning', but of knowledge painstakingly acquired by scientists. Here, once again, is the gulf between humanity and other species.

POISON SHYNESS

Among our behavioural similarities to other species are aversions, especially to poisons. Chapter 4 describes findings, by Dennis Chitty and others, on avoidance of novelty (neophobia). The same group observed another, very different kind of withdrawal—one that arises from an unpleasant experience.

A pile of unfamiliar food on a rats' runway is first avoided: it is a *new object*. (The delay is greater if the food is in an unfamiliar container.) But eventually it is sampled. Afterwards, however, the food is again avoided, as if 'fear' and 'curiosity'

Can Rats be Counted?

A critical reader of this narrative will have noticed unsupported statements about rat numbers. How can one make a census of small, nocturnal, well camouflaged, burrowing animals? The Oxford zoologists (chapter 4) 'counted' wild Norways by feeding them (surplus baiting). Piles of wheat grains soaked in water are put out to induce all the rats in an infested area to adopt the wheat as their main food. The amount consumed then rises daily, to reach a peak after about eight days. This figure (in grams) is divided by 24 (the average daily consumption in grams of an adult Norway), and so gives a minimum figure for the population. (Surplus baiting should not be confused with prebaiting, described below, which requires putting down only small amounts of food at each baiting point, as a preliminary to poisoning.)

Surplus baiting as a method of census can give consistent findings. It is, however, useful only if a single rodent species is present, and no other animals are competing for the food. When I attempted it in subtropical Malta, large ants at once arrived and removed the grains, one by one, to their nests. Infuriating.

An alternative and more difficult procedure, often used in American studies of city Norways, is estimation by traces of the rats' movements: burrows, tracks, droppings, smear marks and others. This requires much experience and skill.

No one method is entirely satisafactory. (For more on the tricky problems of census, see the book by Barnett & Prakash on Indian rodents.)

(or hunger?) are opposed. But the former gradually gives place to unhesitating consumption. If, however, the food contains a poison, such as zinc phosphide (Zn_3P_2, once much used by pest controllers), the small amount first taken causes illness but may not be lethal. The animal then stops eating for a time. When it resumes, as a rule it refuses the toxic mixture: it has become bait shy. It may even reject separate constituents of the mixture with a distinct taste, such as sugar.

Many people have had a similar experience. If we become ill after eating an unusual food, we avoid it afterwards, sometimes for years. Moreover, the aversion develops only when the ill effect is felt, perhaps after some hours.

For rats, the survival value of neophobia, combined with acquired aversions to poisoned foods, was shown in many field experiments. The obvious method of poisoning rats is to offer them a lot of a toxic mixture; but the kill is then often less than 50 per cent and most of the remaining rats have become bait shy. The population that remains can breed at its highest rate and quickly restore its numbers.

One remedy is to train the rats to eat the bait. For several days, small piles of plain food are put down on the runways or in burrow entrances. Most of the rats then become accustomed to the food and, when poison is added, 'ravin down their proper bane'; and die. Populations thoroughly treated in this way may be reduced to below 85 per cent.

The *prebaiting* procedure, however, uses much labour, and the poisons are dangerous. An alternative, developed a little later, was to employ a substance, warfarin, which interferes with blood clotting. (It is also used during surgical operations.) Such anticoagulants have no immediate ill effect and so do not induce an aversion (bait shyness). They are also unlikely to harm human beings or domestic animals. But, after eating the bait for some time, rats die from internal bleeding.

When it was first used in the 1940s, warfarin seemed to be the ideal rat poison. It and other anticoagulants have indeed now largely replaced acute poisons. But another obstacle has arisen: some rats are immune to the effects of clotting agents. Since this difference from other rats is genetically determined, a campaign based on anticoagulants is likely to produce a population of resistant rats. Such populations, first recorded in England, have now emerged in other European countries and the USA.

Worse, as Manuel Berdoy and Pete Smith have described, the genetics of resistance is not simple. At first immunity seemed to be a dominant condition, that is, due to a single dose of one (uncommon) gene; but now several mutant forms of the relevant gene have appeared. Hence resistant populations are not all genetically the same.

A showy way of headlining these difficulties would be to say: 'Nature fights back'. We return to populations and the problems of their management in chapter 9.

PSYCHOLOGISTS DEVELOP AN AVERSION

Bait shyness is an example of *learning after a long delay*, a subject with an odd history. People working in applied science, such as pest control, usually hope to get information from 'pure' science, that is, from researches not primarily directed to solving economic or similar problems. But knowledge of food aversions at first went the other way.

The early findings on the effects of poisons had two disconcerting features. First, in conventional experiments on learning (described in chapter 5), habits are acquired slowly and need many trials; but aversions arise from a single, unpleasant experience.

Second, and more important, in laboratory experiments the interval between stimulus and the animal's response is usually brief—often only a few seconds; but the interval between ingesting a poison and experiencing illness may be some hours. It seemed that, when illness is involved, an immediate impact is not needed for learning.

This was confirmed in elegant laboratory experiments by John Garcia and his colleagues. Their method ensured that illness followed ingestion only after a delay. They gave rats an unfamiliar, agreeably flavoured, harmless mixture, such as saccharin in water; and, after a chosen interval, they induced illness by injecting a poison. At the extreme, the interval could be as long as twelve hours. As predicted, when the rats had recovered, they rejected the novel mixture. Controls were injected in exactly the same way but with a harmless solution. (Another method of inducing illness was to use a small dose of injurious radiation.)

Zoologists (as far as I know) did not find Garcia's findings at all shocking. Certainly, the survival value of learned aversions is obvious enough. It is, after all, the obverse of the ability to choose favourable foods (pages 92–6). But some psychologists found the findings aversive, just as others had recoiled from the idea of latent learning (page 79).

Sam Revusky has described the reluctance of the psychological establishment to accept learning after long delays. Garcia did the first relevant experiments in the 1950s, as a technician who had received no rigorous training in 'learning theory'. (This unbiased state was evidently an advantage, as it had been for the Oxford workers.) But he and others then had much difficulty in getting their papers accepted for publication. Even as late as 1976, the influential weekly journal, *Science*, published an attack by a leading psychologist on the concept of food aversion learning. A rebuttal was submitted but was rejected.

Revusky believes that some of the hostility was due to recent upsets in psychology. The findings of two prominent research programs had been exposed as completely erroneous. (Those on the genetics of intelligence were even believed to be fraudulent.) Nobody wanted to risk another scandal. He also points to the influence of a scientific training in which the dangers of accepting misleading findings are emphasized. It is, he says, just as important to be ready to adopt true ones. But he writes too that his account 'agrees with the hypothesis of many sociologists of science that the unwillingness to accept radical results stems primarily from jealousy and a reaction to threat'.

He might have added that, most of the time, scientists (including psychologists) get on with their daily work without drama or obvious passion. (I have written more about what research is really like in *Science, Myth or Magic?*)

A CLINICAL CONTRIBUTION

Despite these obstacles, our knowledge of human diet and responses to foods owes much to experiments on domestic rodents. I.L. Bernstein and others have described a special case. They wished to reduce the severe nausea and vomiting caused by drugs used to treat cancer; and they began by asking outpatients in a Seattle Hospital to eat some specially flavoured (and enjoyable) ice cream shortly before they were given a nausea-producing drug. Many patients happily agreed. When, however, the ice cream was offered again some weeks later, patients often refused it. They responded rather like rats which had had a single experience of illness after eating a novel food. This was the more notable because, since they were human, and could talk to the doctors and nurses, they knew that the nausea they had experienced was *not* due to the ice cream.

The experimenters then asked whether findings on aversions shown by Norways could suggest ways of reducing human nausea. Bernstein writes:

> Since taste aversion studies in the rat pointed us to drug-aversion studies in humans, it seemed appropriate to turn to animal models to learn more about the mechanisms involved. [Shortened.]

Rats were therefore given cyclophosphamide, a therapeutic drug which causes nausea in people and is also aversive to rats. How, it was asked, could the aversion be prevented? Fasting the animals while giving the drug did not work; nor did giving their water a strange flavour. Success was achieved only when the usual diet was replaced, on the day of treatment, with a strange diet. The eventual outcome was that patients' diets were adjusted to produce some relief from nausea.

Research of this kind demands great tenacity and, often,

a refusal to give in to disappointment when experiments fail. It is also an example of correct method. Animals were used to test conjectures. When a promising finding emerged, it was received with caution and was further tested by studies of people.

SOCIAL FEEDING: BLACK RATS IN PINE FORESTS

Human beings usually learn habits, skills, manners and even many aversions slowly, by social means. We are taught what to do, or at least we imitate others. Imitation, teaching, emulation and encouragement, though immensely complicated, are so familiar that we often expect to find them in other species. But, as a rule, teaching involves elaborate speech, which has no counterpart in animal signals.

Social learning, however, can occur without speech. This has now been shown (or 'taught') to us on a large scale, even by lowly mammals. In Israel, for more than a decade Ron Aisner and Joseph Terkel have studied feeding by black rats in pine forests. The rats rarely come to earth: they nest in the trees (*Pinus halepensis*); feed on the pine seeds; and drink the dew that accumulates at night. The seeds provide a complete diet and are obtained by stripping the covering scales from the cones: first, the scales at the base of a cone are gnawed through, then other scales are systematically removed in a spiral pattern. The behaviour gives an impression of great dexterity and skill (page 107).

A remarkable observation was made in the laboratory. When black rats from other habitats were offered pine cones, they failed to strip them effectively. The newborn young from such 'incompetent' females were fostered, soon after birth, on to females proficient at stripping; and *these* young developed the ability to strip cones. They seemed somehow to have picked up information from their foster mothers. Correspondingly, in control experiments, young born to 'stripper' females were fostered on to nonstripping mothers and did *not* develop the stripping ability.

Evidently, strippers do not differ genetically from nonstrippers. The stripping habit must therefore depend on social transmission, but just what to call it is debatable. Terkel, the leading worker in this field, puts it 'within the teaching category'; but, as he shows, no ground exists for saying that the mothers actively teach their young, or even that the young imitate their mothers. The crucial requirement, evidently, is that when it is beginning to feed on solids, a young rat should experience partly stripped cones: it does so by snatching incompletely opened cones from its mother. The behaviour then comes, as we see below, in the category of stimulus enhancement.

Clearly, more remains to be found out. Terkel asks, for instance, 'How did the phenomenon get started in the first place?'

SOCIAL FEEDING: WHITE RATS IN CAGES

In nineteenth-century writings, we find a persistent story of rats warning (or teaching) others about the dangers of poison bait. No details and little evidence are offered, which is not

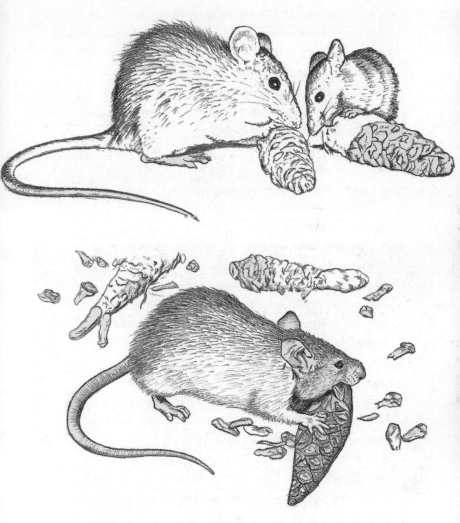

Top: *A young black rat feeds on a partly open cone which it has seized from its mother.*

Bottom: *A black rat strips a pine cone to reveal the edible seeds. (Drawings after Walter Ferguson, courtesy Joseph Terkel)*

The Kinds of Social Learning

All complex human social action is founded on social learning. The main kinds of such learning are listed below. The extent to which other species learn in these ways has been much debated.

'Imitation'

Stimulus enhancement (drawing attention to something)

Emulation ('making a splash')

Imitation in the strict sense

'Teaching'

Encouragement (offering an opportunity)

Deterrence (keeping from danger; driving away)

Teaching in the strict sense

Several kinds of social learning are familiar from watching the behaviour of children. A small child plays with a toy on seeing another playing with it. This is called stimulus enhancement. Or one child may see another achieve something, such as making a splash by throwing an object into water; the first child then does the same thing. This may be called emulation ('I can do it too!).

Casually, we might say that both are examples of imitation; but the word imitation is better reserved for when an individual observes another perform an unusual act and then goes through the same novel sequence. It is, however, often difficult to be sure that this has happened.

Much human learning depends not only on imitation but also on teaching by others: the skilled person *persists in instruction until the pupil achieves a certain level of performance or improvement*. I call this 'teaching in the strict sense'. Social transmission of skills and habits results in local traditions. These are passed on (or 'inherited'), from generation to generation, not by genes but by example. In human societies, tradition is inherent in our social conduct and our skills. Transmission is not only from elders to young: it is often among individuals or groups of the same generation; it may be (and today often is) from young to old. It makes possible rapid change in social practices and in skills. We might call ourselves *Homo docens*, for teaching in the strict sense occurs only among human beings.

(Other classifications of social learning are possible. For more, see reviews by Galef, 1992, and by Heyes, 1993; and one by me, 1994, on teaching as distinctively human.)

surprising: the idea of a mother rat telling her little ones, 'No, no, not those grains: they contain nasty poison', is of course preposterous. Nonetheless, it is reasonable to ask whether a rat's feeding behaviour is affected by an encounter with another rat made ill by poison.

Jeff Galef, his colleagues and others have asked a more general question: to what extent is a Norway's choice of food influenced by the presence or traces of other rats? Despite the differences between the domestic varieties and the wild, they courageously embarked on more than a quarter of a century of meticulous and detailed studies of laboratory Norways. Their findings suggest that domestic rats retain strange abilities which presumably aid the survival of wild rats in freedom.

Wild Norways, we know, are 'central place foragers': they usually live in burrows and, when feeding, they range widely outside them. (They may, however, take food home. This enables them to eat under cover.) When they return home after feeding they evidently smell of the food eaten.

In many of Galef's experiments, a rat is allowed to eat outside its home and then to return to its nest. This 'central place' houses another rat. One question was: would the second rat, given a choice, be influenced *against* the food if the first rat were ill? No such effect was found. Food aversions, it seems, are not acquired by detecting illness in others.

Young rats can, however, be drawn away from toxic foods by the behaviour of adults. When they become independent, they tend to go where other rats are already feeding. This stimulus enhancement (page 108) can perhaps lead wild Norways, in freedom, to eat nourishing rather than harmful food.

Strangely, food aversions can be *cancelled* by a social encounter. In an experiment, a rat first eats a novel, but harmless food. It is then made ill, and duly develops an aversion to the food. It next meets another rat which has been eating the same food, but without suffering any ill effects; and, as a result, it abandons its aversion and eats the food.

What then of positive preferences? In some experiments, two foods were made available, each distinctively flavoured— for instance, with horseradish or cayenne pepper. An animal encounters another rat which has recently eaten one of them, say, horseradish. The question was: would the second rat, given a choice, prefer horseradish to pepper? On the whole, it did. And vice versa.

Evidently, the choice of food of a laboratory Norway can be influenced by the postprandial odour of another. Galef and his dedicated colleagues, however, took nothing for granted. They embarked on extensive further studies, some of which involved sniffing their subjects' breath. They also used a more advanced method of chemical analysis (mass spectrography), which revealed the presence of carbon disulphide, CS_2, in the breath of their rats. Further experiments showed odorous, sulphur-containing substances to influence the preferences of responding rats.

Hence laboratory Norways are equipped to transmit information about food to their neighbours, and to receive it. Recently, similar findings have come from wild Norways kept in a large colony. Manuel Berdoy gave his rats grains flavoured with either peppermint or cinnamon. Both flavours initially repelled the rats. Some individuals, however, were released after being scented with cinnamon behind their ears and elsewhere on their fur. Their odour was therefore like that of a rat which had eaten the flavoured grain. Other members of the colony then ate more cinnamon than peppermint. Evidently, their preference, like that of Galef's domestic Norways, had been socially altered.

EARLY LEARNING

For rats, one odorous channel of communication is milk. The milk of a female Norway is flavoured by the foods she is eating. These can influence the preferences of her young when they begin to eat solids.

Still more remarkable are prenatal influences. Peter Hepper offered pregnant domestic Norways cloves of garlic daily for six days. 'Females', he writes, 'showed no hesitation in eating the garlic.' When their young reached twelve days, and were mobile, they were given a choice between garlic and onion.

Can (Bad) Breath Be a Social Signal?

Do these findings tell us anything about ourselves? For decades, writers of popular science have energetically tried to explain human social action by genes or 'Darwinism'. No ultradarwinian has, however, (to my knowledge) yet proposed that genetically determined breath odours act as human social signals. Perhaps this opportunity was missed because we are dominated in our daily lives by what we see, and tend to disregard personal odours. As well, many readers, reminded of them, will think of those we find offensive. They might even hesitate to talk bluntly about them. Recall the encounter of the great but often unwashed lexicographer, Samuel Johnson (1709–1784), with an acquaintance who exclaimed, 'Dr Johnson, you smell!' He replied, 'No, Madam, you smell; I stink.'

My own favourite literary association is with one of Shakespeare's most lighthearted sonnets, in which he assures us,

My mistress' eyes are nothing like the sun . . .
And from some perfumes is there more delight,
Than in the breath that from my mistress reeks.

In a modern review, R.L. Doty, of the Pennsylvania School of Dentistry, comments on the dearth of information about human social [and antisocial?] odours and on the widespread use of scents to obscure them; but he shows that we can get information about people from their breath—not only bad breath. Five men and five women, in rigorously regulated conditions, inhaled the breath from 33 healthy persons with a high level of 'dental hygiene'. The ten judges, especially the women, scored high in distinguishing the sex of each subject. All judges tended to assess male breath as stronger and less pleasant than female.

The question whether this kind of information is important in our daily lives is, it seems, not yet answered. Nor does this work tell us whether people in other communities would show similar (or greater) abilities.

Great care was taken to ensure that the young were not affected by the odours of the female's breath or body, but all the young showed a clear preference for the garlic. Control young, of females which had not been given garlic, showed no preference.

(That the odorous substance in garlic crosses the placental barrier between mother and foetus is, I am told, well known to midwives in countries where garlic is popular. The first gasp given by the newborn is likely to be strongly flavoured.)

Learning: Limited and Unlimited

Learning after long delays has limitations. Norways, we know, avoid *tastes* associated with illness; but, if they are given a slight, aversive *shock* after drinking a novel, tasty mixture, they do *not* avoid that mixture later. Shock and illness are not associated. Here is a peculiarity which we may plausibly attribute to natural selection. The associations which can be learned are evidently those that aid survival in nature. The ancestors of rats were not exposed to electric shocks.

The reader may remark that nor was electric shock a feature in our ancestors' lives in the Paleolithic. Yet we are capable of making associations impossible for other species. We meet here, once again, the gulf between us and them: we are intelligent enough to have the concept of cause and even to tell jokes about it. One story concerns experiments in which domestic Norways failed to thrive when fed on a famous breakfast cereal. They are said to have done better when given the cereal wrappings instead.

To sum up, variation in how much a rat eats, and what it chooses to eat, has an astonishing range of causes. First, a rat regulates its intake of energy: the quantity eaten depends on the energy content of the food. In this aspect, the rat is a physical system maintained by negative feedbacks. Other internal states influence choice of food, especially when the animal is deficient in salt or a B vitamin. Choice then depends on previous experience of the foods available. More surprising, at all stages of development preference can be influenced by what other rats have been eating.

All this illustrates a fundamental biological principle: the development of an organism depends on a continual interaction with an inconstant environment. Individual behaviour, like all other features, is an outcome of this interaction. As we see further in the next chapter, to understand living things we need to know how to analyse and to apply this principle.

7

ALL IN
THEIR GENES?

That instincts exist on an enormous scale in
the animal kingdom needs no proof.
WILLIAM JAMES, *Principles of Psychology* (1907)

Animals have often been said to survive not by exerting
intelligence but by instinct. In the recent past, theologians
held instincts to be conferred on animals by a benign deity.
Today, some writers substitute genes for instinct and sound
more modern. Even human beings have been described as
machines created by their genes.

In this chapter I ask the reader to shake off such fancies.
We are now concerned with the sorts of things we can truly
learn, especially from experiments on laboratory animals,
about some difficult problems of heredity, environment and
evolution. As illustrations, I take four histories: one on the
genetics of navigation by Norways; one on early influences
on the brain; one on rats as pets; and a fourth on the
'Lamarckian heresy'.

SOME UNNATURAL SELECTION

A central concept of biological science is natural selection. Although sometimes presented as a dogma, it has been critically scrutinised ever since it was proposed. The scrutiny has included experimental tests. Research, however, is rarely directed to grand theories about 'Life, the Universe and Everything': the experiments, like those described in previous chapters, have usually been about details.

In the 1920s, E.C. Tolman, already mentioned in chapter 5, decided to study the genetics of maze learning. In casual

Natural Selection

Natural selection arises from variation. Differences among individuals are universal. The chemistry of heredity, combined with sexual reproduction, tends to maintain this diversity. Some varieties differ genetically and breed more successfully (are 'fitter') than others. One consequence has been familiar for millennia: selection of animals and plants for domestication has given us the animals which live with us on our farms and in our houses. Around 1900, it also produced our tame and tractable domestic Norways (chapter 2).

Hence genetically determined differences in fitness can lead to changes in whole populations. Applying this notion to all living things led to the idea of evolution by natural selection. Small modifications, of kinds observable in a single lifetime, are held to accumulate and so to produce the gigantic evolutionary changes of the past few billion years. The questions we can hope to answer by experiment, however, concern the modifications we can induce during a few generations of selective breeding.

conversation, the ability of rats to find their way about might be attributed to instinct or assumed to be inherited ('genetic'). Tolman asked what seemed to be a fairly simple question: can it be altered by selective breeding?

Though an ebullient character, he refrained from calling the press and announcing an impending breakthrough. Instead, like so many of us, he performed a number of experiments which gave confusing results. He then, very sensibly, handed the problem to an able pupil, R.C. Tryon (1901–1967). As a result, a project was launched which continued, in various hands, for decades. The findings had and have implications both for biological theory and for human biology.

Tryon divided a mixed lot of 142 domestic Norways, variously coloured, into two groups. For eight generations, one group was selectively bred to solve a particular maze problem quickly; the other was bred to solve it slowly or not at all. The maze forced the rats to make many choices on which way to turn.

Both Tryon and the first group of rats were amazingly successful. They still evoked, however, as far as I know, no headlines in the press. Tryon presented his findings in the *Journal of Comparative Psychology*—a learned publication read, if at all, only by specialists. Moreover, he did not say that selection had produced a colony of genius rats, or even that the rapid learners were more intelligent than the dullards. He named the two groups 'maze-bright' and 'maze-dull'.

Yet this admirable objectivity and restraint still did not match the limitations of the experiments. Maze learning is not a simple trait. Domestic Norways, we know, can use not only vision, but also sound and contact to find their way about (chapter 5). The two stocks were eventually tested in a variety of conditions and the 'dull' rats, but not the 'bright', were found to rely largely on vision. Whether they scored as dull or bright depended on the maze and the lighting in which they were tested. In some conditions, the rats originally assessed as 'dull' did better than the 'bright'. The rats bred for 'brightness' had evidently been selected for extra ability to use one of the available senses.

That this could be done even with domestic Norways is an example of a general finding: almost any population of organisms has a reservoir of genetical differences which allow selection to take effect when the environment changes.

In experiments of this kind, the animals are kept in standard laboratory conditions which, as we know, are quite unstimulating (chapter 5). R.M. Cooper & J.P. Zubeck, in Donald Hebb's laboratory, reared rats of each strain in wire cages furnished with only a food box and a source of water. When they were tested as adults, the performance of the 'bright' rats was poor and little different from that of the dullards. Like human children reared in slums, they were dumbed down by their wretched conditions.

Others, however, were reared in an 'enriched' environment which included 'ramps, mirrors, swings, polished walls, marbles, barriers, slides, tunnels, bells, teeter-totters, and springboards'. The 'bright' rats now did little better than in an ordinary environment; but, more important, the performance of the 'dull' rats was much improved.

These findings are a particularly clear example of the fundamental and difficult principle already encountered in the previous chapter: that every trait is an outcome of a long and intricate interaction between what is present at fertilisation of the egg cell and a varying sequence of environments.

PETTED RATS

One environmental feature, important for Norways, is handling. If you pick one up, even very gently, it is likely to stiffen and to extend its legs fully, as if poised to fall. And if you now drop it on to a flat surface, it will indeed land on its feet, unharmed (the righting reflex). But some rats behave differently: when picked up, they lie back, relaxed, rather like a pussycat waiting to be tickled. These rats have been frequently handled and probably stroked, or 'gentled'.

Even wild Norways can be tamed and made to relax while handled. To achieve this, one merely plays with them regularly from an early age. They can then become agreeable pets. (They can also be highly disconcerting to visitors to one's laboratory who are accustomed only to normal Norways which bite vigorously when handled.)

Psychologists were alerted to 'gentling' when experimenters found it to improve the resistance of domestic Norways to various ills. In some conditions, it reduced susceptibility to gastric erosions (a murine equivalent of gastric ulcer).

One outcome was increased care during experiments. If one is comparing two groups of rats, they must each be managed in exactly the same way in every respect, including handling. The need for such caution has been experimentally tested. Students were given distinctively marked rats, taken from two groups, for behavioural experiments. One group,

Stimulating the Synapses

In 1986 an American newspaper published an advertisement with the headline: INCREASE THE SIZE OF YOUR ORGAN. Above it, however, was a picture of the human brain. Readers were invited to improve their intellectual powers by reading the paper.

Does varying the environment produce differences in brain size, and, therefore, behaviour? Until the middle of the twentieth century, the growth of the brain was commonly assumed to be fixed in advance, presumably at fertilisation. Little or no complex interaction of genes with environment was allowed for. Today, we know better.

Some years before that headline appeared, Elsie Widdowson (1906–2000), a nutritionist in Cambridge (England), showed me some nutritionally deprived domestic Norways. They seemed to her, from casual observation, to be exceptionally active. How could this be tested? Here was one of those seemingly simple questions which, to answer, demands a lot of work.

I collaborated with her and a colleague, Jim Smart, in a series of experiments, in which we asked whether an early period of protein deficiency (and the consequent high intake of carbohydrate) influenced adult behaviour. Positive findings might suggest implications for children, many of whom experience just this sort of deprivation.

Our subjects, hooded Norways, were each observed, undisturbed, in a residential maze for twelve days. Long after the end of deprivation, the formerly protein deficient rats were consistently more lively (hyperactive?) than their undeprived cousins. The difference persisted even during a period of food shortage. (They resembled the restless rats deprived of vitamin B1, described in chapter 5.)

Could the effect be reversed? Unfortunately for this project, soon after the experiments were completed I emigrated to Australia, hence they were not followed up. American psychologists, however, especially M.J. Renner & M.R. Rosenzweig (who provided the newspaper headline above), had already begun a lengthy project which compared the effects of 'enriched' (that is, complex) environments with those of the 'impoverished' condition of a cage. Their stimulated Norways tended to have larger brains and to be better at problem solving than the others. In addition, 'enriched' housing *reduced the ill effects on behaviour of early malnourishment.*

The effect on brain weight seems to have been due to increased growth of the processes of the nerve cells (dendrites). These make synaptic contact with many other neurons. This history therefore illustrates once again how, every time one looks, one finds new interactions with the environment during the development of an organism.

G.B. Shaw (1856–1950), in *The Doctor's Dilemma*, has a physician who exclaims, 'Stimulate the phagocytes!' Had Shaw written his play a few decades later, he might have had a character who, with equal justification, could have advocated stimulation of the synapses. This too, of course, is an 'environmentalist' assertion.

they were told, was much more intelligent than the other. Supervisors then unobtrusively recorded the behaviour not of the rats but of the students. The students tended to take more care of the 'intelligent' rats. (The two lots of rats were in fact identical.) The prejudiced response by the students, which could have influenced their findings, was quite unconscious.

When, therefore, one is attempting certain kinds of rigorous research, the experiments should be performed by assistants ignorant of the hypothesis to be tested. 'Theirs not to reason why . . .'.

Another outcome of the work on gentling was its impact in the media. Journalists took it up and headlined the need to cuddle babies and young children. For a time, 'cutaneous stimulation' became a slogan. Some parents responded with amusement: they may have needed expert advice on matters such as diet and illness but, as most of us do, they already hugged their children as a matter of course. They could justly have added that the effects of handling white rats, however rigorously studied, do not justify statements about people.

Those who advocated cutaneous stimulation had an answer. The findings from the laboratory, they agreed, proved nothing about human beings: they merely warned us not to take current customs for granted. The psychologists and paediatricians, and the parents they addressed, were living in a society in which infants and small children were often left on their own for long periods, for instance, in prams; at night even the smallest infants commonly slept alone. They asked whether such practices are desirable; and, the answer was often no.

Observations on animals can then provoke salutary scrutiny of human action. That scrutiny leads to consideration not of genes but of life styles—environmentalism again. It directs our attention to everyday features in the lives of our children, which we can usefully control.

ARE RATS DARWINIAN?

The need for rigour in experimenting emerged with another controversy which can still hit the headlines. Shortly before I began this chapter, a newspaper, stirred by recent research,

Genes and Environments

In the modern media, the expression 'gene for' is often heard. Scientists, we may be told, have found a gene for the ability to make accurate drawings, or for homosexual preference, or for a condition such as osteoporosis. But genes themselves are 'for' the first stage of a long, complex development. They interact with other genes (which also vary) and other cell components and so promote the synthesis of proteins in the cells of the developing or developed organism. These proteins produce structures which themselves interact with a changing environment. A trait, such as an ability, a preference or a disease, is the outcome of interactions which are difficult to describe, let alone explain. Some traits (say, a rat's tail) are very stable in development and therefore appear in all, or nearly all, members of a given species. But it is never appropriate to say that they are *genetically determined*, as if some traits exist which are *not* so determined. All traits are both 'genetical' and 'environmental'. How *variation* in each trait is influenced by genes or conditions of rearing can, as a rule, be discovered only by rigorous experiment.

announced: LAMARCK PUTS EVOLUTION IN A SPIN. An obvious comment is that evolutionary theory has, ever since Darwin, been continually revised and developed; and to ask, has Lamarck helped?

The Chevalier de Lamarck (1744–1829) was one of the great eighteenth-century naturalists. He did important work on classifying plants. Later he classified animals and gave us the distinction between vertebrates and invertebrates. He was possibly the first to use the word, biology. He was also a dissenter, for he did not accept the creation of species described in *Genesis*. Lamarck studied geology and realised that rocks and fossils must have taken millions of years to form. The species of animals and plants, he said, arise by small steps and this takes a long time. He was an evolutionist fifty years before Darwin published his great work, *On the Origin of Species*.

But what marked him off from other naturalists was his proposal on the causes of evolution. His researches on many organisms led him to emphasise adaptation: all species, he saw, are very precisely fitted to their surroundings. Adaptation, he said, was due to the effects of exertion, and of striving for change. He also held that disuse leads to degeneration. A feature of his theory, which is not often mentioned, is the need for adaptive modifications of both sexes.

Changes in the environment continually force organisms to make adaptive changes in their physiology or development. An eminent English zoologist, P.B. Medawar (1915–1987), in a restrained but thorough refutation of Lamarckism, takes the example of thickening of the mammalian skin in response to chafing. This is familiar from the hardened epidermis of the hands due to manual work. It is, of course, a useful effect (and is owed, in a sense, to striving).

In one region of the epidermis, of both rats and human beings, extra growth is already present at birth: the newborn has thickened skin on the sole of the foot. Medawar describes this difference from the skin of other regions as 'developmentally prefabricated': he writes, 'it could not have arisen as an adaptive response *in utero* because the foetus treads water in so far as it treads at all'. Lamarckism, in effect, attributes that thickening to striving, or at least to running on hard ground, by ancestral mammals.

Modern knowledge of genetics and development presents

a different picture. Large organisms begin life as a fertilised egg with a nucleus which contains chromosomes. The material transmitted from one generation to the next consists largely of genes in the chromosomes. *Characteristics* are not inherited, as if they were property passed on by a legacy. During development, the products of gene action combine with environmental effects to produce the traits of the finished organism such as a tree, a mushroom or a human being. During the interaction, the genes usually remain unaltered. Hence they are transmitted unchanged to the next generation.

A modern Lamarckian would therefore be obliged to attribute the thick skin of mammalian feet to changed genes; and the changes should match the improvement in the skin produced by running. The offspring with changed genes would then develop thicker skins than usual, without exertion. The environment has to act on the genes and *change them adaptively,* so that they produce thick skins.

Genes, of course, do alter: rarely, they mutate. The changes are not directed by need; in that sense, mutation is random. What we know of the material basis of heredity, therefore, in itself seems to rule out Lamarckian effects.

Nonetheless, in science, if somebody proposes a hypothesis, however improbable, a next step is to test it. A crude experiment was done a century ago: the tails of newborn Norways were cut off, generation after generation, to see whether eventually they would give rise to tailless young. Of course, they did not. But this hardly tested Lamarck: there was no striving by the rats, only by the experimenter; and we are not told what happened to his children.

Other experiments have put real demands on the subjects.

A psychologist, W.M. McDougall (1871–1938), English but of Scottish descent, studied the effects of learning a simple task. Rats had to escape from a water tank with two ways out. One was brightly lit; when they chose it, they received a mild shock. The other, dimly lit, let them painlessly out. Each kind of exit was sometimes on the left, sometimes on the right.

Rats were trained for many generations; and successive generations improved in the time they took to escape without shock. It seemed that a Lamarckian transmission of an acquired ability had occurred—that is, a memory had been inherited.

This sensational finding led to violent controversy. McDougall's experimental methods were minutely examined and flaws were found. W.E. Agar and others, in the University of Melbourne, repeated the experiment: they bred rats for twenty years and 50 generations; and, for the first ten years, *they achieved the same result*. But they took an essential precaution. They ran a group of exactly similar rats at exactly the same time but did *not* train them. And these controls improved too, over a similar time. Later, the performance of both groups declined, and then improved once again. So, whatever caused the improvement, it was not the training.

The fluctuations, though never fully explained, were attributed to environmental features which could not be held constant. One was the rats' food; another was a seasonal influence on the rats' performance—perhaps the temperature of the water in which they swam to their goal.

This project, kept going with unfailing tenacity during two decades of depression and world war, does not show how evolutionary change happens, but it does tell us much about the demands of biological research and about the biological variation one meets during experiments over long periods. The findings on rats match those on other animals; in fact all the well designed experiments on 'Lamarckism' tell the same story: they disconfirm Lamarck's ideas about evolutionary change.

Yet Lamarck was right to emphasise the influence of its surroundings on the developing organism. For him, the adaptive features of an organism were the outcome of striving to cope with the environment. He lived, however, before biological experimenting took off and long before the science of heredity was founded. The influence of the early environment, shown so clearly in the work on 'bright' and 'dull' rats, delivers a warning. Changes in complex traits have no simple explanations: they are due to the interaction of many varying genes with many inconstant features of the environment. The latter may be difficult to identify even in a highly simplified laboratory situation. This principle is reinforced when we turn to social behaviour.

8

RAT SOCIETIES

What checks the natural tendency of each
species to increase in number is most obscure.
CHARLES DARWIN (1859)

Here is a condensed version, consisting of verbatim passages, first published in 1966, from an account of the social lives of wild Norways and black rats.

These animals display collective aggression of one community against another. Toward members of their own group they are models of social virtue, but they change into horrible brutes when they meet members of other colonies. Pair formation can, however, occur between members of different groups. Members of pairs then jointly attack others, prevent formation of new pairs and so exert a tyranny over them. A female may creep up on another, leap and bite the side of the neck and so induce fatal bleeding. If a stranger approaches a rat in its territory, members of the whole colony are alerted and display group hatred: black rats communicate the alarm by a shrill, satanic cry. With their eyes bulging from their sockets, the rats then set out on a rat hunt. Intruders are slowly torn to pieces.

Every statement in this farrago is incorrect or misleading. The source is chapter 10 of a once widely read, popular work, *On Aggression*. The author, K.Z. Lorenz, here seems to have been—in a famous phrase— intoxicated by the exuberance of his own verbosity. He provides an extreme example of crediting animals with regrettable human qualities. The compulsion to draw conclusions from supposed likenesses, and the longing for exciting stories—often ludicrous yet enjoyable—is so general that such anthropomorphism might qualify as a natural human instinct.

The paragraphs above are also an example of the prominence of conflict in descriptions of animal behaviour. The preoccupation with violence matches the crude 'Darwinism' of the catch phrase, 'the survival of the fittest'. (It also goes with a longstanding and justified anxiety about human violence.)

In serious studies of animal behaviour, two phenomena, both often discussed under the heading of aggression, have been, and still are, prominent: status among members of a group, and the territories held by neighbouring groups or individuals. The names, *status system* (or *dominance hierarchy*) and *territory*, are derived from human social action. So are the terms, *dominance* and *subordinacy*, used to describe status relationships. They are, nonetheless, important in all authentic accounts of animal societies.

SOCIAL SIGNALS: THE SHOW OF VIOLENCE

The principal founder of modern ethology (the science of animal behaviour) was Nikolaas Tinbergen (1907–1988). He

Animal Signals Classified by 'Meaning'

1. 'Look out!': alarm signals.

2. 'Go away!': deterrent signals occur in territorial and dominance interactions and when a female weans her young.

3. 'I love you!': prolonged and elaborate courtship is widespread.

4. 'I'm friendly!': an animal may signal, and so avoid a clash.

5. 'Help!': distress calls are heard when a young bird or mammal is in need of food or warmth.

6. 'Come here!': a parent may signal to the young.

7. 'This way to food!': ants, bees and many others guide members of their colony, especially by odour trail.

was much concerned with status and territory, but he did not emphasise violence. On the contrary, in 1951 he wrote:

> ... it is a very striking and important fact that 'fighting' in animals usually consists of threatening or bluff. Considering ... that sexual fighting takes such an enormous amount of the time of so many species, it is certainly astonishing that ... physical struggle is so seldom observed.

Like genetics and experimental psychology, Tinbergen's ethology began, unavoidably, with simple situations. Emphasis was on objective descriptions, usually of social behaviour, sometimes supported by experiment. Standard signals were identified for each species studied: for instance, a flashing light from a mating firefly or a patch of colour on the beak of a parental gull. The signals were called 'releasers', as if they were triggers, and were described as innate (or instinctive). Each was said to represent a distinct state of the signaller and each evoked a 'fixed action pattern' or standard response, typical of the species, from another animal. In their crudest form, they resembled those of an automatic railway: green, and a machine moves; red, and it stops.

To what extent, I asked many years ago, does this scenario match the social signals of rats? Since a nocturnal, burrowing species cannot be easily watched in freedom, interactions were staged in large observation cages furnished with nest boxes. They eventually allowed the typical postures of five species of rats to be recorded in detail (fully described in a paper written with Ivan Fox & Wayne Hocking). As we now see, they proved not to match the original concept of the social signal.

Rats are contact animals. They sleep together, often piled in a heap. When males meet, even if they are strangers, one is likely to crawl under the other. Sometimes they crawl over each other. (The females of some species do the same.) Perhaps this distinctive performance deters attack, but that is little more than a guess. Another usually peaceful act is grooming (strictly, allogrooming): while huddling, one rat nibbles at the fur of another. I originally described this as amicable, but later came to have doubts. In close encounters, allogrooming by bush rats (*Rattus fuscipes*) ranges from nibbling to holding the other rat down while tearing out chunks of hair. Nonetheless, all these interactions might come under the heading of social sedation.

An additional enigmatic act, seen during meetings between males of all species, I rashly called the threat posture, or TP (rashly, because 'threat' suggests an intention to punish or hurt, but we have no convincing means of deciding what a rat *intends*). The back is arched, the legs

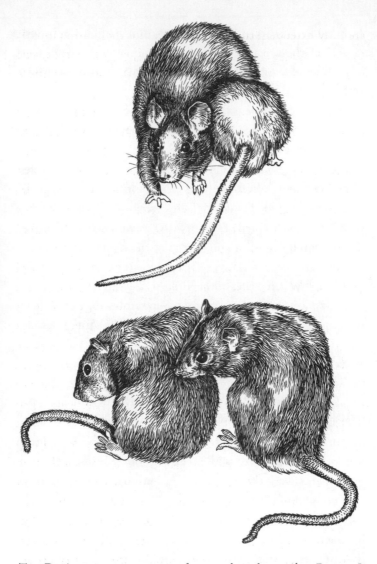

Top: During encounters, one rat often crawls under another. Bottom: In another form of contact, one rat grooms the other. (Drawn by Gabriel Donald from photographs)

are fully extended, the hairs are erect and the head is usually turned toward the opponent. It often precedes attack, and sometimes also comes after it; or two rats may perform it together.

The most exciting interaction is attack, in which one rat leaps on to another with rapid adductions of the forelimbs (which can be seen only in speeded up cinefilm). Sometimes it is accompanied by a brief bite. Rarely, an attacking Norway, or a long-haired rat (*R. villosissimus*), bites and holds on. Attacks are not relentless but are often interrupted: the actors turn to 'boxing' or to other attitudes, especially TP. During these intervals, an attacked rat may approach the attacker.

This complex of fluctuating postures is what the rat sees and feels. But interactions do not depend only on vision. Rats regularly sniff other rats; and, as we know, Norways influence each other by food odours (chapter 6). Wild rats of all species also possess an array of glands which secrete pheromones, of which more than twenty have been identified. Some species have odorous sebaceous glands on their bellies; black rats have cheek glands which they smear on surfaces. Objects are often also marked with urine.

Scent marking by mammals was at first assumed to be a defence of a territory—another instance of supposing social behaviour to be predominantly combative. Later, many scent marks were shown to be attractive. Some observers, however, believe that wild Norways attack other males only if they possess a strange odour. Certainly, an attacking male urinates and defecates as it approaches an intruder. But what rats smell during a clash is still uncertain.

A complete account of rat interactions must also include

The 'threat posture' is common to all the species of Rattus *studied, including the little known swamp rat* (R. lutreolus) *shown here.*

sounds. All the species so far studied make similar noises. A hostile encounter between Norways may begin with percussion—chattering of teeth by the attacker. Whether this is a social signal is doubtful. The sounds uttered by bush rats (*R. fuscipes*) of both sexes have been recorded in detail. During encounters, while they perform the postures described above they also utter pure whistles, harsh screams and intermediates between them. During attack and boxing, both rats scream and whistle but, when one approaches or 'threatens' another, only the animal approached sounds off.

If a reader met a neighbour who whistled and screamed, jumped up and down, emitted powerful odours, kicked the reader's shins and intermittently broke off and made affectionate

Top: Attack by a male Norway. (Drawn by Gabriel Donald from a photograph)

Bottom: Body odours are important social signals. (Drawn by Gabriel Donald from a photograph)

This posture, although called boxing, is nonviolent. Two bush rats.

cuddly contact, I suspect that the reader would feel confused, to say the least. Yet rats faced with such behaviour can respond in ways which allow crowding and a high fertility.

CLOSE ENCOUNTERS

Certainly, these interactions create difficulties for ethologists. Male wild Norways, trapped from crowds, are sometimes found to be scarred, evidently owing to bites by other rats. They might therefore be suspected of continuous strife. Yet, if several adults are put together, all at once, in a large cage or enclosure, with plenty of food and nest sites, they grow and look sleek. So do black rats. When I first did this, I was impressed by their playful behaviour: they wrestled and pounced but did no harm. Later, in groups of both sexes, they mated and bred in those cages, and their lively young went through mock fights. They were fun to watch.

Manuel Berdoy has confirmed the pacific capacities of wild Norways. His group of twenty adult males, kept in a spacious enclosure, all evidently survived unharmed. He recorded relationships, of the kind sometimes called a peck order, in which A defeats B; B defeats C and so on, to form a 'linear dominance hierarchy'. The measure of dominance was success in 'aggressive interactions'.

ASSAULT WITH BATTERY

Clashes among wild rats can be experimentally analysed by introducing a strange male into the living space of another. For maximum effect, the resident should have female companionship. The reader might therefore suppose that males fight *for* females. Not so: in encounters in small colonies, a female Norway in oestrus is followed by several males, which take turns to copulate. Similarly, in Manuel Berdoy's large colony, receptive females 'were assiduously followed by a string of up to 10 males, and mated repeatedly with several of them'. Females not in oestrus are ignored: rats do not form pairs.

In experimental clashes, a female need not be present during the actual encounter. Some of those I staged were with a single resident male while the female was shut in a nest box. The visitor typically made the first approach, but lively action was nearly always begun by the resident: the newcomer fended off the attacker or ran away. The encounters therefore hardly rated as fights, for they were onesided.

When intruders were introduced to groups of wild Norways

Aggro

The responses of a resident rat to an intruder on its territory are often called aggressive. In ordinary speech, 'aggression' signifies ungoverned violence, or unprovoked assault intended to cause injury. But defence of an occupied area is hardly unprovoked; nor, as we know, may we say what an animal intends.

Calling *defence* 'aggressive' is an instance of putting a strangely mixed bag of activities under this one heading. These activities include hunting by carnivores; territorial displays such as bird song or the grimaces of a monkey in a tree; and sounds and gestures involved in dominance within a group. In such accounts, animals are often described as though they were human and human beings as animals: territorial defence may be equated with the many forms of human property ownership, and the intolerant activities of animals likened to our many forms of war.

As a result, animals and human beings may be held to have an aggressive drive which impels them to attack members of their own species. Sometimes, the drive is said to build up internally if it is not expressed. The tendency to eat or drink necessarily increases with deprivation; but intolerant behaviour is appropriate only in particular social situations. Correspondingly, if an adult male wild rat is deprived of encounters with strange males, it does not turn on members of its own group. Each of the many activities called aggressive—predation, territorial interactions, dominance and others—needs separate analysis, preferably by experiment. In such analysis, imagined drives do not help.

of both sexes, no group action occurred. Three kinds of adult male emerged. 'Alphas' were always large, moved about freely and initiated attacks on intruders: small rats never overcame much larger ones. 'Betas' adapted themselves to an inferior role: they kept away from the alphas (it is tempting to call their behaviour ingratiating); but they fed well and gained weight. They attacked intruders only if the alphas were first removed from the colony. This enabled them to switch to an 'alpha' status. Both alphas and betas appeared also during experiments with the Australian longhaired rat (*R. villosissimus*).

RAT SLAUGHTER

Last were the 'omegas'. Although attacks were intermittent and brief, after a day or two under attack some rats were marked by decline in weight, slow movements, hesitant eating, drooping posture, bedraggled appearance and, eventually, death. Similar debility resulted from encounters between black rats. In human terms, the omegas seemed dejected or seriously depressed.

Such deaths have sometimes been attributed to killing by other rats. I have seen several hundred clashes between wild rats of five species, some quite vigorous; but in none was a rat *killed* by another rat. Correspondingly, Bob Blanchard and his colleagues counted the bites inflicted by black rats in a long series of encounters in Hawaii. Nearly 90 per cent were on the back or tail, where they could do little harm; all were only skin deep; none was lethal. Similar observations have been made on the little Polynesian rat (*R. exulans*) in New Zealand.

Yet even a completely unwounded omega was likely to collapse and die, sometimes quickly but more often after hours or days of increasing debility. I have called this 'death of unknown origin', or DUO, by analogy with the physician's 'pyrexia of unknown origin', or PUO.

The species, other than the Norway, in which DUO has been most thoroughly studied, is the tree shrew (*Tupaia belangeri*) of southeast Asia. Dietrich von Holst, in the University of Bayreuth in Germany, staged clashes like those described above and made observations astonishingly similar to those on Norways. Males attacked intruding males and (in a difference from Norways) females attacked strange females. The attacks were brief and the wounds, at worst, were superficial scratches. Yet the intruder crouched in a corner, in a bedraggled state, except for hurried visits to food or water. To quote von Holst:

> They even tolerated the infrequent attacks of the winners
> without any attempts to defend themselves or to flee. They
> ceased grooming ... and their fur became rough and dirty.
> To the human observer they gave an apathetic or depressive
> impression.

Even if not attacked again, the victim became steadily weaker and, unless removed, eventually died.

In some encounters, however, the response of one shrew (called a subdominant) resembled that of the betas observed among Norways. 'Slight fights' led to dominance relationships and mutual tolerance. The winners grew well and often ignored their opponent. Subdominant animals also grew but were more active, watched the dominant and avoided it by giving way or fleeing. Occasionally, in a clash, they defended themselves.

A tree shrew during a hostile encounter. The raised hairs on the tail reflect a state of arousal. (Courtesy Dietrich von Holst)

The mating of tree shrews differs greatly from that of Norways, for in freedom the adults usually live in pairs, each in a territory. In captivity, although they can be kept alone and healthy for years, von Holst found formation of a pair to improve wellbeing. But pair formation was precarious. Putting a male and a female together sometimes provoked 'intensive fights' and death. Only about 20 per cent of pairings led to immediately amicable behaviour which, in von Holst's words, conveyed a strong impression of 'love at first sight'.

DUO has been closely observed only in captivity, but

unexplained deaths have been reported in free populations. Jack Calhoun introduced wild Norways into a large population in Baltimore; some quickly emigrated but many died. He also saw debility and death in a spacious paddock in which some rats, to get food, had to run the gauntlet of others. H.-J. Telle similarly recorded clashes among Norways and black rats in Germany. Rats released in an occupied region were driven from the residents' burrows and some died without visible external or internal injury.

A DIVERSITY OF SPECIES

The genus *Rattus* includes about 300 species. Although, as we know, the social signals of those so far studied are much the same, one cannot safely predict the conduct of one species by studying others. From what I initially observed of wild Norways, I assumed violence to be a prerogative of males. Later, R.F. Ewer was working in a hut (which she describes as 'officially rat-proofed') near Accra (Ghana). The roof was regularly visited by agile black rats which were very different from Norways. Dominance relationships, violent behaviour and death occurred prominently among females.

Later still, my colleagues and I found the female of another species, the Australian longhaired rat, to be as deadly as the male: she attacks other females but, under attack, may become debilitated. Strangely, two other free-living Australian species, the bush rat and the swamp rat, when similarly subjected to 'social stress', did not succumb to DUO.

Yet a rodent of a different family, that agreeable pet the

The Pacifist Norway—a Boojum?

An esteemed physiological psychologist, Frank Beach, once denounced the use of domestic Norways in research on behaviour: the title of his article, 'The Snark was a Boojum' (inspired by Lewis Carroll), implied that comparative psychology had softly and silently vanished away owing to the preoccupation of experimenters with albino rats.

The descriptions in the preceding pages are of wild rats and so perhaps escape this reproach. All the *domestic* Norways, white, hooded, even brown or black, that I have studied, were very different: they were uniformly pacific. During decades of selection in captivity they seemed to have completely lost some components of wild type social interaction. They could therefore be caged and moved around without regard to their companions.

Another feature, occasionally observed, was indifference to being attacked: I have seen a wild male Norway energetically 'threaten' and pounce on an albino, while the albino drifted around a large cage taking no notice of the attacker.

In 1950, N.L. Munn, a leading experimental psychologist who knew only domestic Norways, wrote:

> Relatively little research has been done on social behavior
> in rats, primarily because rats are not especially influenced
> by each other's actions ... One observes in rats none of
> the complexities of behavioral interaction found in higher
> mammals.

As a result, I became quite snobbish about the superior social repertoire of wild Norways. I had, however, failed to allow for variation among the many strains: some greet strangers with merely a sniff, but others wrestle vigorously.

More important, social interactions depend greatly on group structure and the environment. N. Adams & Bob Boice gave a colony of domestic rats more space than is usual in a laboratory; and, after some weeks, individuals resembling wild alphas, betas and even omegas were observed. Similarly, R.J. Blanchard and his colleagues have induced energetic conflict among domestic Norways by keeping them in groups for many weeks and regularly introducing strangers.

Once again we meet the need to take account of variation in development in a variety of possible environments. Genetical differences are often crucial but, as usual, they explain only part of the differences we see.

Effects of Domesticating Norways

	wild	domestic
Response to strange object in familiar place	avoidance	approach
Response to strange place	exploration	exploration
Response of adult male in territory to intruder	attack	sniff, huddle
Males, with females, in artificial colonies		
mortality	high	low or none
growth	may lose weight	all grow
movement of males	one moves freely	all move freely
Exceptions exist to all these rules.		

guinea pig or cavy (*Cavia aperea*), can do so. Norbert Sachser and his colleagues have kept cavies in stable groups for many years. The males in such groups form harmless dominance relationships with each other and pair with individual females.

But strange males are attacked. Some intruders fight back (but always lose). Others merely submit or flee; and these may die in a few days, sometimes even if removed from the group.

Encounters were also staged between pairs of males like those, described above, between Norways. Each resident male was accompanied by a female. The outcome depended on the upbringing of the intruders. Those reared in socially stable groups could adapt to a subdominant (or beta) status. Others, each reared alone with a female, collapsed; but none was wounded. Hence death of social origin is emerging as a widespread but unpredictable and still obscure phenomenon; of which more in chapter 10.

◆

The postures and screeches described above are, as far as is known, common to all species of the genus *Rattus*. They are signals of the kind called species-typical (formerly, innate). This implies standardisation; but they are quite unlike the fixed action patterns described at the beginning of the chapter. Some ethologists call such encounters 'negotiations'; but this word suggests the use of language and is hardly appropriate. Together with odours, they form a fluctuating complex which evokes a similarly varying range of possible responses. They remind us that biological systems are rarely like the machines designed and made by human beings. The conduct of two members of the same species during an encounter is predictable, if at all, only as a probability; and the outcome of a clash is still less certain. Incessant variation is a feature of living systems of which we see more in the next chapters.

PART III

THE BLINDNESS OF RESEARCH

To achieve convincing explanations of the social interactions of wild rats, voles, cavies and others, we must do experiments. At least four methods can lead to experimental tests: those of ecology, ethology, genetics and physiology. They are often, and necessarily (for life is short), used alone. Hence their practitioners may seem like the legendary blind men trying to interpret an elephant: one seizes the tail and reports a rope; another touches an ear and identifies a large leaf...

But this is unfair, especially to those who have crossed the boundaries between disciplines. One impressive body of findings is due to ecologists. They began by counting their animals; but later they made a fruitful marriage of ecology with population genetics and a second, bigamous one with ethology. Their work shows us yet again the inconstancy of the material on which biologists have to work: diverse species; each species genetically various; those differences themselves varying over generations; a fluctuating physical and living environment to which individuals respond in a variety of ways; a fluctuating social environment; and, of course, the waywardness of human action.

9
POPULATION EXPLOSIONS

Destruction reigns. There is dismay ... and
demands to Authority. Authority remembers
its experts or appoints some: they ought to
know. The experts advise a Cure. The Cure can
be almost anything: golden mice, holy water
from Mecca, a Government Commission, a
culture of bacteria, poison, prayers ... a trap,
a Pied Piper. The Cures have one thing in
common: with a little patience they always
work. They have never been known entirely
to fail. Likewise they have never been known
to prevent the next outbreak.

CHARLES ELTON, *Voles, Mice and Lemmings*

While I was writing chapter 6, the human population
of India was announced as having passed one billion
(one thousand million). No statement was made on the number
of rats or other rodent pests: with good reason, for—in India
as elsewhere—nobody has counted them or could even make
a reliable estimate of their numbers. Despite or because of
their strange social interactions, rats have a fabulous capacity

Maternal behaviour: a Norway with a young litter.

to multiply: and the members of some species can live crowded in colonies of many hundreds.

Rats of the species most studied have no breeding season: female Norways can come into oestrus at any time. Gestation is only 21 or 22 days, and the female becomes receptive again about 21 hours after the birth of her young. She may have as many as ten young in a litter. If she is inseminated while she is feeding a litter, her next gestation is lengthened by delayed implantation of the embryos; but this has only a minor effect on her formidable fertility.

150

The 'Maternal Instinct'

A breeding female performs activities which are not only highly effective: they also seem to be both predictable and firmly fixed. Like some social signals (page 130) they have been called innate or instinctive. Prominent are nest building and bringing back straying young to the nest. A pregnant female reliably builds a nest from the materials available—plant fragments or, in captivity, paper strips or cotton—and retrieves her young to it. Such behaviour seems compelled by an internal drive (or instinct), and to develop without learning by practice.

But it is not. Females have been reared in cages without anything which could be picked up or manipulated. Their food was powdered and even their faeces fell through a floor of netting. At maturity, they were mated in ordinary cages and produced litters. But they did not make nests: instead they scattered nesting material around their cages. They similarly failed to retrieve their young to a single place, but carried them around without obvious direction. The deficient behaviour resulted from an environment which did not offer the usual opportunities to learn how to manipulate objects. (Compare the findings on maze learning in chapter 5.)

The maternal behaviour of the rat provides examples of activities which are notably *stable in development* (a phrase owed to Robert Hinde); but, as these findings show, they develop not in a vacuum but in an inconstant environment. Once again, as in previous chapters, to understand behaviour we have to consider the dynamics of development from egg to adult.

DENSITY DEPENDENCE

What encourages the breeding of wild rats? More important, since it always stops eventually, what stops it?

The hordes of rats in human communities are an outcome of human action: we grow and store great concentrations of food; we construct shelter in buildings and drains and at the edges of fields; and we often kill the carnivorous species which prey on small mammals. Moreover, we usually kill the rats themselves only when they are numerous. The slaughter may then be impressive, but the likely result is a surviving population which can breed at a high rate.

Eventually, however, the increase always declines. This suggests an adverse influence, of which the effect increases with population density. If we could identify such density related factors, we could learn much about how populations are regulated. But the idea of density regulation is misleadingly simple. The perfect case is an isolated population in constant conditions: the curve of increase in numbers is then S-shaped: increase is slow at first; becomes rapid; then slows and stops.

A big gulf exists between such a formal curve of population growth and the numbers in a rat-infested padi field, warehouse or sewer. If the curve represented reality, most populations should be steady at the top of the curve; yet nearly all fluctuate, if only because the weather does. Worse, they interact with populations of other species which, too, are varying. These others include predators such as snakes, hawks and foxes. But it is often difficult to decide whether the predators are controlling the prey or the prey are determining the numbers of predators. The sewers of New York City are reputed to have a thriving community of alligators, owed to the escape or release of pets. These voracious predators are said to live on the rats. But rat numbers remain high and, evidently, well able to support a population of muricidal reptiles. Unfortunately, the sewers of a large city are not easy to study as an ecosystem.

Granted, as we see below, sometimes altering a single factor, such as the food supply or shelter, can have a distinct effect on numbers; but, as a rule, no one density related factor can be identified as the key to keeping numbers down.

Possible Checks on Numbers

- Shortage of food or water
- Lack of shelter
- Predators
- Pathogens (microbes or parasites)
- Social interactions

UNDERGROUND WARFARE

Among the amenities we lavish on rats and other pests are shelters. A population may exhaust this protection before it runs out of food. Chapter 1 describes the massive populations of rodents in the bunds which border Indian rice fields. Even reducing the height of these banks reduces the numbers of rats, gerbils and others. Similarly, if mud walls in villages are replaced by hedges of cactus (*Opuntia*), one important kind of shelter disappears.

Some human structures are mobile. Ships were notorious for harbouring thriving populations of rats, especially *Rattus rattus*. Two centuries ago, a Tasmanian newspaper described the numbers of rats coming ashore from a convict ship as beyond belief. These creatures were almost as far as they could

The Irony of Chernobyl

In 1986, an accident in a nuclear reactor at Chernobyl in the Ukraine led to the release of a cloud of radioactive dust and calamitous fires. All 135 000 people within 30 km of the powerhouse had to leave their homes. The emptied region became a wilderness. When writers of science fiction describe the outcome of such an event, the picture is often of bare rock, or a shattered surface inhabited by a few hideously distorted, mutant organisms.

More than a decade after the accident, biologists described the actual scene. Wild life is flourishing. Among the mammals are badgers and otters in the waterways (which carry plenty of fish); several species of deer in the woods; and predators such as wolves and lynx. Uncommon birds breeding there include a crane, a stork and a sea eagle; some belong to endangered species. The whole region is now proposed as a permanent nature reserve.

The only species known to have declined are commensals deprived of the facilities offered by humanity. Of them, the most notable are, of course, rats.

possibly be from their port of origin. Their descendants are probably still with us.

Far from the rice fields and the former colonies are the sewers of European and other cities. These are crucial for human survival in thronging urban environments, but they are a disaster for pest management: if a Norway could design an ideal environment for the survival of its species, it might well propose a system of sewers and drains.

My own introduction to economic zoology came through the London sewers. In the 1940s, a graduate from Oxford, I

Research in the London sewers. A well protected technician collects bait from a specially installed tray. (British Ministry of Food, by permission)

was directed into the wartime Ministry of Food. I could confidently describe the strange, massive skulls of the Cotylosauria (a long extinct group of primitive reptiles) and much else of a similar sort; but of living mammals I knew little, and of public health, nothing at all.

London at the time was believed to be under threat of overwhelming bombing. The health authorities had visions of massive destruction not only of buildings but of sewers, and the release into a shattered city of hordes of voracious, dangerously infected rodents.

At a crowded meeting of medical officers of health and chief sanitary inspectors from the London boroughs, I was faced with questions on what should be done about sewer rats. My answers were necessarily brief. But fortunately, the Oxford zoologists, whose work I knew, had begun to provide some solutions (chapter 4). Hence, at every accessible place in this enormous system of tunnels, prebait was put down; later, poison was added. Vast numbers of rats must have been killed. How much disease was prevented and how much food saved can only be guessed.

AWKWARD QUESTIONS

Later, I was sufficiently foolhardy to draw attention to the necessary guesswork. The problem arises especially when the objective is to protect food. One tactless question I asked was therefore this.

> If, in a given area, one hundred men are employed to use current methods of pest destruction, will those men save more food than they would produce if they were farm workers? Or should there be fifty men, or two hundred?

Eventually, some of us tried to find out what happened when rat populations were 'controlled'. As expected, if a perfunctory poisoning was carried out in a sewer system, some rats were killed and numbers were quickly restored. But a thorough operation of two poisonings, each preceded by prebaiting, could reduce a population to a small fraction of its original size. Recovery was then slow, but only if infestations at the surface were also dealt with.

To be sure of the effect of an operation, one needs an area with well defined boundaries. After the war, one became available in the south-west of England, where the military authorities had left a rural region empty of soldiers but full of wild Norways. The human inhabitants of a nearby infested village were, understandably, annoyed. So we tried to exterminate the rats. We were unsuccessful but, a year after a double prebaiting and poisoning, the population was still only a third of its former size.

One implication concerned the policy of the local authorities. Each infested farm was treated with poison four times a year. Better and more cost effective results would have been achieved by thorough double poisonings annually. The problem was then to persuade administrators to change their practices and the farming community to accept new arrangements.

Meanwhile, in the United States, parallel experiments had been done in urban environments. Baltimore, then as now (page 47), had formidable populations of wild Norways which were studied by J.T. Emlen and his colleagues and by D.E. Davis. In a typical experiment, the rats of a large residential block were estimated to number a hundred. All (it was believed) were killed. Rats reappeared in small numbers, but only after 17 months. They were allowed to breed and, after a further six months, an explosive increase led to a population larger than before. Eventually, with many fluctuations, it seemingly returned to around the original figure. Again we see the need to hit a population when it is down, instead of waiting until it has bounced back and done a lot of damage.

RICE RATS AND THE ECONOMIC THRESHOLD

Many of the exertions described above arose from the anxieties of global war and its aftermath. The principles then developed are now being rediscovered, especially by ecologists and others working in Asian countries. In these regions, with their many pest species, estimations of the numbers of rats, and of the ultimate effects of attempted control, are rarely more than guesses.

An exception is in the work of B.J. Wood on the rice rat (*Rattus argentiventer*) in Malaysia. These rats have been described as the most important rodent pest in south-east Asia. They not only eat large amounts of rice but also cut down the rice plants at the base, after the fields have been drained. The destruction they cause resembles that due to mole rats (*Bandicota bengalensis*) in India and elsewhere (page 16). Yet their biology is still not fully known and even knowledge of their diet is incomplete.

Wood used cubes of maize mixed with a waxy substance and laced with the anticoagulant warfarin (page 100). He compared large treated areas (of about 40 hectares) with others left alone, and achieved a substantial increase in yield in the treated areas. The cost of the campaign was trifling compared to the value of the extra crop produced. Hence in this case attempts at controlling pests were not too expensive to be worthwhile. A useful slogan to express this principle is now available: the *economic threshold*.

Later work in south-east Asia has been summed up by A.P. Buckle. Rat poisons, including anticoagulants, are much the same as they had been half a century earlier. Unfortunately, the anticoagulants, so important for controlling Norways, are slow to take effect against other species; and even against Norways they have limitations (chapter 6). Buckle writes:

> Few who devise and evaluate rodent management strategies fail to advocate integrated approaches. [Such] programs must be environmentally sound, cost-effective, sustainable, capable of application over large areas and recognisably advantageous both for growers who implement them and politicians who ... fund them.

But he adds, 'All too often those who conduct rodent control programs pay only lip service to integrated pest management ... and rely almost solely on rodenticides'.

New methods are needed. Meanwhile, all the known ways of discouraging rodent pests should be put into effect simultaneously. One of the most effective is to change the conditions we offer them. The effects of such changes are not simple. Reduced cover and shortage of nesting material obviously make construction of nests and rearing young more difficult, but they also alter behaviour. We need to know not only about the rats' food and shelter but also about their social lives.

SOCIAL SEDATION VERSUS SOCIAL STRESS

Despite human exertions, no species of rat is known to be threatened with extinction. All those I mention in this book

Social sedation among Norways. Another example of the importance of contact.

are alarmingly successful partly because they are versatile: each species seems to be adaptable to diverse environments.

This adaptability is reflected in their elaborate social interactions, described in the previous chapter. Accounts of rat communities (including my own) commonly emphasise clashes. Yet obvious features of groups of Norways and black rats are contact, crowding and fecundity. Despite the reality of conflict, a complete picture should include the balance between social sedation and social stress. During the struggle to understand population dynamics, the favourable effects of crowding, or at least of contact, have been little acknowledged; which is strange, for it is a prominent feature of human experience.

As we know, behaviour sometimes switches from cosy contact to intolerance and dispersion. What effect has their capacity for conflict on the performers? Charles Elton, one of the founders of modern animal ecology, suggested as early as 1942 that 'social antagonism' among crowded rodents could control numbers. At first a simple hypothesis seemed tenable:

the animals could be likened to molecules in a gas which collide more often as the density rises; the more collisions, the more deaths or the fewer young.

But encounters are not all the same: some are sexual or otherwise 'amicable'. We are, moreover, faced with the incoherence of intolerant encounters. The apparent ambivalence of acts such as crawling under and the 'threat posture' has hardly been discussed, let alone explained. Yet it seems to have considerable survival value, for it is found in all the species studied.

Although we do not know what induces apparently sedative behaviour at one moment, seemingly uncompromising hostility at the next, we can identify what provokes vigorous clashes. Male Norways (and others) energetically attack strange males which intrude on their living space. (The females of some species also attack intruding females; but why only of those species and not others is quite obscure.)

One clue therefore seems to be their territorial behaviour. Exclusive occupation of a living space is widespread in the animal kingdom: even marine worms may defend a cleft or burrow against others of their species. But territories are diverse. Among mammals, they range from a lactating female's nest to a region of many hectares.

Territorial behaviour spaces a population out. It may be the principal means of dispersion for animals living in burrows and visiting scattered food sources. In an experiment by J.B. Calhoun some of the Norways in a large enclosed population were molested by others whose territories were nearer the only source of food. The persecuted rats grew less well than the others and were less fertile. (The experiment did not allow

In a Calcutta godown, Indian mole rats assemble in crowds to feed on the stored rice. The large individuals are bandicoot rats. (Courtesy S.G. Frantz)

emigration. In a more usual situation, some rats would likely move away to other sites.)

The obverse is seen when territorial spacing is prevented (unintentionally) by human action. Crowding, measured by numbers to an area or volume, is influenced by the amount of food available. When the supply is lavish, population densities are reached much above those in other environments. An American zoologist, J.J. Spillett, has reckoned that Calcutta godowns house 78 mole rats to every hundred square metres of floor space. This very high figure suggests that, in those conditions, territorial interactions were in abeyance. Here, evidently, is an instance of density related factors interacting.

Plenty of other complexities are on offer. Social effects on population growth need not entail conflict. Among the effects of crowding is delay in the development of sexual maturity in young females. Researches on many species also show crowding to lower fertility. Correspondingly, reducing a population leads to an increase in the production of young. In one experiment, energetic trapping removed about half the Norways in a residential area of Baltimore. The pregnancy rate of the remaining rats rose by about 60 per cent.

The histories of research in this and the previous chapters each describe the encounters of research workers with increasing complexity. As our knowledge grows, so does our awareness of questions for which we have few conclusive answers. Most baffling is the final conundrum of this book: that of death.

10
SOCIAL LIFE
AND DEATH

The characteristic of a living object is that it
reacts to external stimuli rather than being
passively propelled by them. An organism's life
consists of constant mid-course corrections.

R.C. LEWONTIN, *The Triple Helix*

To see a recently healthy animal collapse and die, without
evident cause, is disconcerting. And, on reflection, it
becomes worse, for it does not make sense. When I first
studied this phenomenon, I gave a paper to the Physiological
Society in London. Afterwards, I asked some medical
physiologists what could have caused my rats' deaths. They
tended to mutter something about vagal syncope and to sidle
away. This is much the same as saying, 'heart failure'. Physicians
talk of vagal syncope when the heart stops while full of blood.
But heart failure is *caused*; or so one must assume.

Fieldworkers at that time, faced with similarly obscure
mortality, resorted to talk of 'shock disease'—in effect,
acknowledging that they were baffled.

Other suggestions have been 'psychological', for instance,

humiliation or hopelessness. In 1957, Curt Richter, an American psychologist of great distinction, made an elaborate likening of unexplained rat mortality with what he called 'voodoo death'. But practices such as voodooism depend on speech: the deaths, if they really occur, are due to curses, especially those uttered by a witch doctor. The idea of bewitchment entails symbolism, as when a model of an enemy is stuck with pins to induce injury, or when fragments of hair or nails are held to give power over a person. People believe in these magical influences because they have been taught to do so. Richter's proposal, therefore, likens a uniquely human practice to a strange feature of animal life and death, and so seems to be an extreme case of anthropomorphism.

THE SEARCH FOR INSTABILITY

Today, if a human being dies unexpectedly, one usually consults not a witch doctor but a pathologist; which is what I did. In 1960, a thorough post mortem was carried out on a series of wild Norways, each of which had reached the point of death owing to social persecution. Nothing was found in any of them to explain their condition.

Where should one look next? Perhaps in classical physiology. In 1878, at a time of war and revolution, Claude Bernard (1813–1878), the son of a French peasant, had founded modern physiology with his *Leçons sur les phénomènes de la vie*. It contained the first full statement of the principle of homeostasis. A free life, he wrote, depends on internal stability.

The word 'homeostasis' was brought into physiology much

later, by the American physiologist, W.B. Cannon. His researches in the first years of the twentieth century were an extraordinary anticipation of the times in which they were published. The first edition of his *Bodily Changes in Pain, Hunger, Fear and Rage* was dated 1915, the second year of the First World War. While armies of young men were 'discovering their adrenals' in the mud and death of the Western Front, Cannon described the response of the mammalian body to an emergency demanding battle or flight.

Cannon also wrote of stress. But this concept became prominent in physiology only in the 1940s, the period of the Second World War, through the work of the Canadian physician, Hans Selye. By 'stress' Selye meant the bodily response to any agents which upset homeostasis. He proposed a bold, all embracing theory of a single pattern of responses, called the 'general adaptation syndrome', to a variety of ills. The latter, called stressors, included infection, cold exposure, burns and wounding. His book, *The Stress of Life*, is dedicated to readers who realise that they can 'enjoy the stress of a full life' only with 'intellectual effort'.

Since then the concept of stress has become more complicated. Different stressors are now known to have different effects. Also, individuals respond differently to the same stressor (compare the beta and omega rats, and the subdominant and submissive tree shrews, in chapter 8).

Moreover, the external conditions which provoke the stress response are not always unfavourable: effects of stressors on behaviour include arousal, raised alertness and heightened attention. For a human being, these may be welcome. (Recall the compulsion to explore and the need for novelty described in chapter 5). Also familiar are the exciting effects of vigorous sports, which may entail pain. These activities, however, are usually enjoyed only if they are willingly undertaken: painful exertions are unlikely to be accepted if they cannot be escaped. Correspondingly, the ill effects of being attacked are shown especially by animals which, like Pavlov's 'neurotic' dogs (page 74), cannot get away. (For more, see Constantine Stratakis & George Chrousos.)

MORE SOCIAL STRESS

To investigate socially caused death, it was necessary first to think of questions which could be answered. One, based on elementary physiology, was quickly obvious. During clashes, rats, tree shrews and others use much energy. Perhaps this could lead to low blood sugar (hypoglycaemia), which can lead to collapse.

The blood sugar of enfeebled wild Norways was estimated and found to be high; indeed, perhaps higher than usual. Similar observations were made by other workers on tree shrews and field voles (*Microtus*). Sometimes, the livers of these animals had a low reserve of glycogen, that is, of the store from which extra blood sugar is made when needed. This is typical of an animal adapting effectively to exertion.

The raised blood sugar depends on the adrenals. Many afflictions (or stressors) provoke extra adrenal activity, including exposure to cold and infection. Correspondingly, when rats or tree shrews come into conflict, their adrenals enlarge. The presence of even an inactive member of the same species can

The Adrenals and the Immune System

The adrenals (or suprarenals) nestle against the kidneys. The inner part of each gland, the medulla, originates in the embryo as part of the sympathetic system—the part of the nervous system which responds quickly to alarm or sudden pain: the changes it evokes include raised heart rate, dilation of the pupils, pallor from reduction of the blood supply to the skin and raised hair.

The medulla secretes two hormones: adrenaline (epinephrine) and noradrenaline (norepinephrine). These too make part of the response to a situation demanding 'fight or flight' (in Cannon's famous phrase). They should therefore play a leading part in the physiology of conflict.

The adrenal outer layers, which form the cortex, originate in an embryonic structure which also gives rise to the gonads. They secrete a number of hormones of the same chemical family (steroids) as the reproductive hormones (of which testosterone, oestrogen and progesterone are familiar). During the response to stressors, one group of cortical hormones (the glucocorticoids) act with the medullary hormones to increase the level of glucose in the blood.

The steroid hormones also interact with the immune system. The white cells (leucocytes) in the blood, of which there are several kinds, are the principal source of resistance to infection. They (and the red cells) are produced in the bone marrow in balanced proportions. In some diseases, the proportions change and white cells produce antibodies which attack the invading organisms. The endocrine and immune systems are now known to be in equilibrium; but in severe stress glucocorticoids can *decrease* the immune response to infection and other stressors. This is perhaps a negative feedback which prevents an excessive immune response. It may account for the flaring of latent infections during social stress.

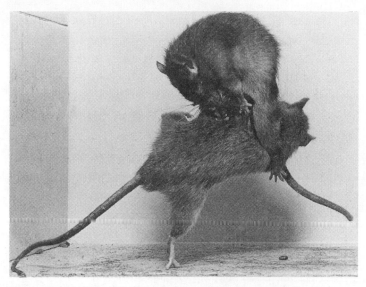

Encounters may use a lot of energy, as in this performance by two black rats.

have this effect. But here is a paradox. The adrenal response is adaptive: it helps to prepare the animal for exertion. What are we to make of the animals which give up, collapse and die? One hypothesis proposed was that death was due to excessive production of adrenal hormones; but no evidence of this could be found.

Meanwhile, wild rodent populations were being searched for signs of infective illness, and many showed signs of kidney disease. In these populations the rates of hostile clashes rose as numbers increased. At the same time the incidence of infection went up too. But this was to be expected: when animals (or people) are crowded together, infection is likely to rise because the chances of transmission are enhanced; and the virulence of a disease organism may be increased by rapid transmission among its victims. These findings therefore did not tell us anything conclusive about unexplained death.

In the laboratory, however, one can control for effects of crowding. Studies by Dietrich von Holst revealed kidney disease, which sometimes proved lethal, in caged tree shrews under attack. This was not due to crowding. Similarly, in Australia we found renal pathology (glomerulonephritis) in longhaired rats socially stressed in controlled conditions in captivity. If, then, a bacterial infection contributes to DUO, it should be possible to prevent it. In experiments carried out with M.H.R. Sandford, a powerful antibiotic, neoterramycin, was given to socially stressed black rats; as predicted, the only deaths were among control rats which received no antibiotic.

These findings forced attention to the ways in which mammals resist infection (page 168). Knowledge of the immune system has increased explosively in recent decades, and von Holst has shown the immune system of a tree shrew to respond to social stress as if it were challenged by an infection. After some days of confrontation with an opponent, his dominant animals (attackers) were growing well and showed little change from their normal condition. The subordinate animals were more agitated than the dominants and showed some immune depression, but this, as we know, did not prevent adjustment to their new circumstances. The most marked changes were shown by defeated animals. Von Holst writes: 'In submissive animals, substantial changes were … found in those types of leucocytes [lymphocytes and eosinophil granulocytes] indicative of strong immune suppression.'

Is an immunological account of DUO possible? Perhaps. One hypothesis concerns soluble proteins, called cytokines. These are secreted by white blood cells especially when the cells are resisting infection. Cytokines also stimulate the secretion of adrenal hormones. One, interferon, has had exposure in the media, for it was found to prevent multiplication of pathogenic viruses. Cytokines—and this is the point—also have a strange 'side effect': they induce loss of appetite and lethargy and so may contribute to the collapse of omega or submissive animals. If so, they provide an example of a response which is usually adaptive but can be fatal.

IS DUO DARWINIAN?

Many questions remain unanswered. In biology it is usual to ask of a trait: how does it contribute to survival? Yet to ask this of DUO sounds ludicrous: can death have survival value? This question implies an assumption which is often made but is nonetheless wrong: that every feature of an organism is a direct result of natural selection, and that it therefore aids survival or breeding. But, as Charles Darwin himself knew, many traits are only indirect consequences of natural selection: they go incidentally with advantageous features and do not themselves aid survival. This, correlated variation, is of great importance, above all for the human species. As Richard Lewontin has recently written: 'Many features of organisms are the epiphenomenal consequences of developmental changes or the functionless leftovers from remote ancestors.'

So the physiological picture of DUO is still incomplete.

- The most obvious anomaly is that death can occur without wounding or other obvious cause.
- Most physiological changes found during the collapse of omegas or submissives resist the effects of stressors. They should therefore prevent death.
- The postures of lethal encounters resemble those seen also during harmless interactions ('play').
- Betas evidently possess special features, not yet identified, which enable them to adapt to attack.

The philosopher K.R. Popper, in his *Objective Knowledge*, quotes a scientist's comment on scientific research: 'Our whole problem is to make the mistakes as fast as possible'. Popper writes:

> The scientist's trials and errors consist of hypotheses. He formulates them in words, and ... can then try to find flaws in any one of these hypotheses, by testing it experimentally, helped by his fellow scientists who will be delighted if they can find a flaw.

As Popper also adds, every solution to a problem leads to new problems. This clearly applies to the struggles with death described above. Much has been learned; but the deaths due to social stress remain an enigma.

11
NATURE AND HUMAN NATURE

Admittedly, we all try hard to avoid error; and we ought to be sad if we have made a mistake. Yet ... if we do not dare to tackle problems which are so difficult that error is almost unavoidable, then there will be no growth of knowledge.

K.R. POPPER, *Objective Knowledge*

Chapter 1 begins with a fifteenth-century bishop's curses. Today the science of living things enables us to do many things unimaginable in his time: every chapter of this book describes achievements owed to the exertions of people ranging from naturalists to mathematicians. Hence current popular accounts of biological science consist largely of confident findings and buoyant conclusions. So they should. To end my last main chapter with questions rather than answers may therefore seem perverse.

Yet the biology presented to students and the public is only a distorted outline. Often, the science of life is held to be reducible to physics. This presumption had a beginning with the French philosopher and mathematician, René

Reminders about the Nature of Life

Here, to avoid vague waffle, is a list of the perfectly familiar features of living things. If they were also features of molecules, atoms or ultimate particles, physicists and chemists would be decidedly upset.

- Each member of a species differs, if only slightly, from all the others in appearance, physiology or behaviour. The differences may be genetically or environmentally influenced; commonly, they are both.

- However stable it may seem to be, even an adult organism is continually changing: its stability depends on a flow of matter and energy regulated by an intricate network [a metaphor] of negative feedbacks.

- A complex organism usually develops from a fertilised egg which looks nothing like the final stage. Individual diversity arises during a regulated development (ontogeny) which entails an immensely complex, continual interaction with a changing environment. It ends with death.

- The activities of an organism themselves change the environment. Sometimes the results—for example, burrows and nests—are obvious; but every organism is continually extracting material from its surroundings and contributing other material.

- Each kind of organism reproduces itself, usually sexually. This often requires elaborate interactions, such as those of pollination (a form of symbiosis), or of courtship and parental care. Hence the dead are replaced.

- Last are evolutionary changes (phylogeny) which are, for the most part, too slow to be observed. But experiments have shown something of what natural selection can do in the short term.

Descartes (1546–1650). Descartes was a contemporary of William Harvey (1578–1657), discoverer of the circulation of the blood. In the ancient world the heart had been held to be the site of the mind or soul. Now it became a pump. Correspondingly, for Descartes animal bodies were automata, analogous to machines made by people. The leading American geneticist, R.C. Lewontin, writes:

> the entire body of modern science rests on Descartes's metaphor of the world as a machine. While we cannot dispense with metaphor in thinking about nature, there is a great risk of confusing the metaphor with the thing of real interest. We cease to see the world *as if* it were *like* a machine and take it to *be* a machine.

The physical sciences have given humanity many answers about entities from atoms to planets, most of which are stable over immensely long periods. Even unstable elements, such as radium, 'decay' at a predictable rate. Decay here is a metaphor derived from biology: organic decay is quite different and much more complicated. The word 'decay' in atomic physics is an example of the omnipresence of verbal imagery and metaphor in scientific descriptions.

Granted, some features of living things can be usefully described and even explained in physical terms: bones are levers; eyes are optical systems; muscles convert chemical energy into mechanical energy. But living systems are not machines or anything like them. Nor are they like the units studied in the physical sciences.

Modern knowledge of biology not only answers questions but also obliges us to face new problems. All the units studied by biologists change incessantly. A disease may be 'conquered', with great benefit to humanity, yet reappear in a new form. Each of the pests of which I write has typical features, but none can be relied upon to respond consistently to human action; nor can the domesticated varieties which we sometimes hope will provide clues to our own nature. A central theme of this book is therefore the complexity of living systems. It shows how advances are made by making simple hypotheses and then by replacing them when their limitations become obvious. An ending with questions is therefore a better portrayal of biological science than any cocky statement implying that we know all the answers.

Homo sapiens is the extreme case of a species creating its environment. We do this partly without calculation, but partly on the basis of rational planning after argument. This demands knowledge. The problems of biology are infinite. The most general implication of the present narrative is the urgency of the need to learn more about nature, including our own. Two and a half millennia ago, the Greek philosopher Heraclitus wrote, 'Nature loves to hide'. Today, Nature still hides; and so, all too often, does human nature.

GLOSSARY

adaptation   (biology) Has two important meanings. 1. Physiological (ontogenetic) adaptation is a change in a single individual which enables it to cope better with its environment: for instance, acquiring resistance to disease; enlargement of muscles with exercise; or learning the way about. 2. Genetical or phylogenetic adaptation occurs when a population changes genetically and so increases the chances of survival of its members and their descendants.

aggression   (ethology) A word confusingly applied to many kinds of violence, even to 'threats', and to defensive territorial signals such as bird song. Not recommended for serious descriptions of animal or human behaviour.

allogrooming   Grooming the integument of another animal.

anthropomorphism   Describing the behaviour of animals as though they were human. (Has other meanings.)

antibody   A protein produced in response to the presence of an injurious substance (an antigen, *q.v.*) in the body. A defence mechanism especially important for resistance to infection by bacteria or viruses. See also **immune response**.

antigen   A substance which provokes formation of an antibody.

bacterium   (pl. **bacteria**) Bacteria are microorganisms, nearly always single cells, without nuclei (prokaryotes). They are essential in soil formation and in maintaining the cycles of carbon, nitrogen and other elements. Many are parasites; some cause human and animal disease.

behaviourism The doctrine that the only proper subject of scientific psychology is behaviour.

cell A unit of living matter, usually microscopic, bounded by a very thin plasma membrane and in plants also by a wall of cellulose. A cell usually contains a nucleus (*q.v.*) but the smallest organisms (bacteria, *q.v.*, and others) are cells without nuclei. The largest cells are very yolky ova such as those of birds.

chromosome A body, usually thread-like, consisting mainly of DNA (*q.v.*) and protein. Chromosomes make up most of the contents of the nucleus (*q.v.*) of a plant or animal cell (*q.v.*). In the somatic cells of animals, that is, cells which are not gametes (*q.v.*), and in the cells of plant sporophytes, chromosomes occur in pairs (the diploid condition); the members of a pair are identical in appearance under an ordinary (light) microscope and are said to be homologous. Gametes have only one chromosome of each pair (the haploid condition). The number of chromosomes varies with the species. Chromosomes are easily seen only during nuclear division.

cognition Thinking something out; making use of separate items of information to solve a problem.

commensal Members of different species living in close association but not in symbiosis or parasitism; literally, 'sharing the same table'. Examples are the rats and mice associated with human beings.

conditional reflex A response or act, elicited by a previously indifferent stimulus, as a result of the repeated occurrence of the indifferent stimulus at about the same time as an existing stimulus for a similar act. The classical example is from Pavlov's dogs: a buzzer is sounded before food is offered; at first, the buzzer attracts attention but has no effect on salivation; but, after many repetitions, salivation occurs in response to the buzzer before food appears.

density-related factor (ecology) Anything that tends to reduce the growth of a population and acts with increasing strength as population density rises: includes food supply, predators, disease organisms, shelter and crowding. See also **feedback**.

deoxyribonucleic acid   See **DNA**.

DNA   Deoxyribonucleic acid. A giant molecule consisting of many nucleotides forming a chain; usually two chains are joined, parallel to each other, and coiled in a helix. Each nucleotide contains one of four bases (thymine, cytosine, adenine or guanine) and a sugar, deoxyribose. Found mainly in the chromosomes (*q.v.*) of animals and plants, and in corresponding structures of bacteria (*q.v.*). The order in which the bases occur in the chain is the 'genetic-code' which, with the intervention of RNA, determines the synthesis of amino acids and so of proteins. DNA is reproduced whenever a nucleus divides and is the material basis of biological inheritance. See also **gene**.

dominance   (ethology) See **status system**.

dominance hierarchy   (ethology) See **status system**.

drive   An unidentified or hypothetical internal state that causes a particular activity. Eating, for instance, may be attributed to a hunger drive. Such statements are now being replaced by descriptions of observed behaviour and, when known, of the underlying physiology.

ecology   The science of the relationships of organisms with their environment, including the environment provided by other organisms. An ecologist may, for instance, study associations of species, as in a rock pool, a patch of soil, a desert, a wood or the human skin or gut; the changing numbers and density of a population of organisms of a single species; or the flow of matter and energy through an association of species.

endocrine organ   An organ which secretes hormones (*q.v.*).

epigenesis   The appearance of new structures during individual development (ontogeny); this entails an interaction of the effects of genes with environmental influences. Opposed to the obsolete doctrine of preformation, according to which the organism is already fully formed in the fertilised egg or even in a gamete (*q.v.*).

epizoötic A widespread infection in a population of animals. The zoological equivalent of an epidemic.

ethology The science of animal behaviour. A major division of biological science, like ecology (*q.v.*). In some writings the term is used in a narrower sense—the study of the behaviour of animals in their natural surroundings.

evolution, organic The descent of organisms from very different organisms in the past, traceable in the fossil remains of organisms. Evolutionary change is still going on; but is slow and can be observed in a human lifetime only on the smallest scale, if at all.

exploration In ethology (*q.v.*), movements of an animal about its living space which are independent of any special need, for instance, for food. They represent a tendency to approach strange objects and places and are a means by which an animal learns about its surroundings (exploratory or latent learning).

extinction (behavioural) The decline in performance of an act or habit as a result of its repeated evocation without subsequent reward, that is, without payoff.

feedback The transfer of output to input, in such a way as to modify the input. In negative feedback the effect is to reduce the input; in positive, the input is enhanced. Negative feedback is universal in living systems, from the regulation of metabolism (chemical changes) within cells to the control of the growth of populations.

fitness In biology, the fitness (also known as Darwinian fitness) of an organism is some measure of its contribution to later generations. There is no necessary connection with athletic prowess.

gamete A cell, usually a large egg (ovum, female) or a small sperm (spermatozoön, male), which can fuse with another cell of the opposite sex to produce a zygote (usually a fertilised egg) which can develop into a new organism.

gene Originally meant a unit of heredity of which the existence was inferred from breeding experiments. The units are passed on, unaltered, from parent to offspring. Soon after 1900 (when Mendel's findings were rediscovered), the genes or hereditary factors were shown to be arranged in line in the chromosomes (*q.v.*) of the cell nucleus (*q.v.*). Later, the material of heredity was found to be DNA (*q.v.*). Hence today the word gene usually means a length of DNA. The nucleotides (*q.v.*) in the DNA molecule are arranged in a linear sequence. Each group of three nucleotides (codon) codes for an amino acid. A gene consists of many similar codons. Much chromosomal DNA is, however, 'noncoding'.

genetical (also **genetic**) Usually means 'related to the action of genes'. May properly refer to *differences* between organisms. some differences are genetically determined. Characteristics or traits are sometimes said to be genetical but this usage is not recommended: *all* traits are influenced by both the genes *and* the environment.

genetics The science of variation and heredity.

grooming See **allogrooming**.

homeostasis Maintenance of a steady internal state: for instance, temperature or blood composition.

hormone An organic substance secreted in very small amounts in one part of an organism and carried to another part where it influences metabolism, growth or other processes.

immune response A response to infection which protects an organism and leaves it less susceptible to the infective agent. See also **antibody, lymphocyte**.

in 'In' is often used when 'by', 'of' or 'among' would be appropriate. The result is sometimes odd, even comic. A reference may be made to 'an observation in Donaldson (1979)', when the observation is *by* Donaldson. An excellent book has the expression, 'hoarding in animals' in the title, when, of course, nobody is hoarding anything in the animals: the animals are doing the hoarding. My favourite example is 'Freezing and storage of human semen in 50 healthy medical students'.

instinct 1. In everyday speech, often means the same as intuition or unconscious skill: as when a person is said to avoid a blow instinctively or to understand the attitude of another person by instinct. 2. When an animal performs a complex act without learning how to do it, this may be called instinctive behaviour. In ethology (*q.v.*), this usage is being given up. 3. A third meaning is the same as, or similar to, that of drive or the impulsion to act in certain ways. In some writings expressions like 'hunger drive', 'aggressive drive' and so on still occur. This usage, too, is being given up: instead, the actual behaviour is described and, when known, the physiology underlying the behaviour.

Lamarckism In ordinary speech, the statement that acquired characters are inherited. It is still sometimes assumed that the effects of use and disuse of organs (such as muscles), and of practice, are transmitted from one generation to the next. On this view, apparent skills—such as those of a bee building a comb or a predator tracking and killing prey, are inherited memories, derived from ancestors which had to learn these skills. We have no evidence that such transmission can happen. Our knowledge of the way in which DNA (*q.v.*) operates contradicts it.

latent learning See **exploration**.

leucocyte White blood cell.

lymphocyte A type of small white blood cell which produces antibodies.

mutation A sudden change in the DNA (*q.v.*), usually of a chromosome (*q.v.*). Mutations can occur in any cell nuclei. Those that occur in a gamete produce mutant DNA which can be passed on to later generations. Of these, the most important are changes in single genes (or codons). DNA is very stable: mutation is a rare event. It is, however, speeded up by radiation and by some poisons. Most mutant genes have an adverse effect on the organism; but some, especially in a changing environment, may be advantageous. See also **natural selection; nucleus**.

natural selection A process in nature which results from the existence of genetically determined variation among organisms. As a result of this variation, some organisms contribute more to later generations than do others. They are then said to have greater fitness (*q.v.*) than the others. Evolutionary change is held to be largely due to differences of fitness in this sense. But natural selection can also *prevent* change. See also **stabilising selection**.

negative feedback See **feedback**.

neophilia Approaching or seeking new or less familiar objects or experiences. See also **exploration**.

neophobia Avoiding unfamiliar objects in a familiar place.

neuron A nerve cell has processes projecting from it, often very many. The cells and their processes conduct impulses among themselves; or from sense organs to a central nervous sytem; or from the central nervous system to muscles or glands. Transfer of impulses from one neuron to another is at junctions called synapses (*q.v.*).

new object reaction See **neophobia**.

nucleus The part of a cell which contains the chromosomes (*q.v.*). Present in nearly all the cells of many-celled organisms.

peck order A term for a kind of status system (ethology, *q.v.*).

pheromone An odorous substance which acts as a social signal.

placenta The afterbirth. An organ consisting of both embryonic and maternal tissues, by which an embryo is supplied with food and relieved of waste products. Animals which reproduce in this way are called viviparous.

predation Hunting, killing and (usually) eating members of a different species.

pyrexia Fever.

reduction, explanatory The findings of one kind of study, such as genetics, can sometimes be partly explained by reduction to those of another, such as biochemistry. So knowledge of the chemistry of DNA (*q.v.*) helps us to understand the phenomena of heredity. Another kind of example is the attempt to explain human societies in terms of genes and the presumed past action of natural selection (*q.v.*). The power of reduction, especially in the physical sciences, has led some people to believe that all biology can be explained by physics and chemistry or even in terms of mechanisms. But, if organisms and human beings are to be understood, it is first necessary to make nonreductionist statements about organisms and human beings. These cannot be replaced by biochemical, genetical or other reductionist statements.

species   The smallest unit of classification (or taxon, *q.v.*) commonly used. All the members of a population assigned to one species, if they reproduce sexually, are assumed to be capable of interbreeding. When two kinds of organisms are assigned to different species, it is assumed that they *cannot* interbreed or that, if they do, they will not produce fertile offspring.

stabilising selection   If an organism is very different from the typical or normal for its species, it is likely to be less fit, in the biological sense (see **fitness**), than those nearer the norm. But see **natural selection**.

status system   (ethology) In animal behaviour, a social system which includes relationships of dominance and subordinacy. The expression 'dominance hierarchy' is often used with the same meaning. The exact meanings of dominance and subordinacy depend on the observer and the species studied: typically, a dominant individual has priority of place, and for food or a mate. Subordinates give way.

subordinacy   (ethology) See **status system**.

synapse   The functional connection of two nerve cells (or neurons, *q.v.*). 'Connection' does not signify continuity: a gap always remains between the surface membranes of the two cells. One cell influences another by secreting substances which pass across the gap. Most nerve cells are in synaptic relationship with many others.

taxon   (pl. **taxa**) A unit of biological classification. The principal taxa are, from largest to smallest: kingdom, phylum, class, order, family, genus, species.

teaching In the present book, the word teaching means an activity which alters the behaviour of a member of the same species (the pupil) and tends to be persisted in until the pupil reaches a certain standard of performance or improvement. (In ordinary usage, 'teaching' has many other meanings.) The systematic teaching of skills is a distinctively human trait.

territory (ethology) In animal behaviour, a region, occupied by an individual or a group of animals, from which other members of the species are excluded. Often reserved for a region *defended* from others. Different species have different sizes and kinds of territory. Territorial behaviour should be distinguished from behaviour which maintains a status system (*q.v.*) *within* a group of animals. It should also not be muddled up with human ownership of property.

vector A living carrier of infection, that is, of a disease organism.

virus Submicroscopic particle, consisting mainly of DNA (*q.v.*) or RNA, which infects an organism and may cause disease. Viruses can multiply only in a living cell.

zoomorphism Explaining human social action by likening people to animals.

NOTES ON SOURCES

The notes below list major sources of the findings described in this book. The references that follow give also others mentioned, under authors' names, in the text. Some titles have been shortened.

Research on the behaviour and ecology of rats up to the early 1970s is reviewed in my *The Rat: A Study in Behavior*. Much recent work, especially the economic aspects, is summarised in a volume edited by G.R. Singleton and others.

For the history of the names of Norways and black rats, see A.H. Barrett-Hamilton & M.A.C. Hilton. For the classification of rodents, see J.R. Ellerman.

Plague and other infections are described by N.G. Gratz, W.H. McNeill, J. Nohl and P. Ziegler.

On exploratory behaviour see especially the volume edited by J. Archer & L. Birke and the review by E. Save.

On cognitive ethology, see the volume edited by C.A. Ristau.

Early work on rat genetics has been massively surveyed by R. Robinson. For the principles of modern genetics, consult D.T. Suzuki and others.

Modern studies of rodent pest management, especially in Asian countries (but not India), are reviewed in the volume edited by G.R. Singleton. For the Indian scene consult S.A. Barnett & I. Prakash.

Recent work on social stress among mammals has been thoroughly reviewed by D. von Holst.

An outstanding, recent short review of what biology is really like is *The Triple Helix* by R.C. Lewontin. My *Science, Myth or Magic?* attempts the same thing, but is less technical.

REFERENCES

Agar W.E. *et al.* 1954 Fourth (final) report on a test of McDougall's Lamarckian experiment. *Journal of Experimental Biology* 31, 308–21

Aisner R. & Terkel J. 1992 Ontogeny of pine cone opening in the black rat. *Animal Behaviour* 44, 327–36

Archer J. & Birke L. (ed.) 1983 *Exploration in Animals and Humans*, Wokingham: van Nostrand

Bammer G., Barnett S.A. & Marples T.G. 1988 Responses to novelty by the Australian swamp rat. *Australian Mammalogy* 11, 63–6

Barnett S.A. 1948 Rat control in a plague outbreak in Malta. *Journal of Hygiene* 46, 10–18

Barnett S.A. 1975 *The Rat: A Study in Behavior*, 2nd edn, Chicago: University of Chicago Press

Barnett S.A. 1994 Humanity as *Homo docens*: the teaching species. *Interdisciplinary Science Reviews* 19, 166–74

Barnett S.A. 2000 *Science, Myth or Magic?: A Struggle for Existence*, Sydney: Allen & Unwin.

Barnett S.A., Brown V.A. & Caton H. 1983 The theory of biology and education of biologists. *Studies in Higher Education* 8, 23–32

Barnett S.A., Fox I.A. & Hocking W.E. 1982 Social postures of five species of *Rattus*. *Australian Journal of Zoology* 30, 581–601

Barnett S.A., Hocking W.E., Munro K.M.H. & Walker K.Z. 1975 Socially induced renal pathology of captive wild rats (*Rattus villosissimus*). *Aggressive Behavior* 1, 123–33

References

Barnett S.A. & Prakash I. 1975 *Rodents of Economic Importance in India*, London: Heinemann; New Delhi: Arnold-Heinemann.

Barnett S.A. & Sandford M.H.R. 1982 Decrement in 'social stress' among wild *Rattus rattus* treated with antibiotic. *Physiology and Behavior* 28, 483–7

Barnett S.A., Smart J.L. & Widdowson E.M. 1970 Early nutrition and the activity and feeding of rats. *Developmental Psychology* 4, 1–15

Barnett S.A. & Stewart A.P. 1975 Audible signals during intolerant behaviour of *Rattus fuscipes*. *Australian Journal of Zoology* 23, 103–12

Barrett-Hamilton A.H. & Hinton M.A.C. 1912–20 *A History of British Mammals*, London: Gurney & Jackson

Beach F.A. 1950 The Snark was a Boojum. *American Psychologist* 5, 115–24

Beer C.G. 1991 From folk psychology to cognitive ethology. In: *Cognitive Ethology*, ed. C.A. Ristau, Hillsdale, NJ: Erlbaum, pp.19–33

Bell T. 1837 *A History of British Quadrupeds*, London: van Voorst

Berdoy M. 1994 Making decisions in the wild. In: *Behavioural Aspects of Feeding*, ed. B.G. Galef *et al.*, Chur, Switzerland: Harwood Academic Publishers, pp. 289–313

Berdoy M. & Smith P. 1993 Arms race and rat race. *Terre et vie* 48, 215–28

Bernstein I.L. 1985 Learned food aversions in the progression of cancer. In: *Experimental Assessments of Conditioned Food Aversions*, ed. N.S. Braveman & P. Bronstein, New York: Academy of Sciences, pp. 365–80

Blanchard R.J. *et al.* 1975 Conspecific aggression in the laboratory rat. *Journal of Comparative and Physiological Psychology* 89, 1204–9

Blanchard R.J. *et al.* 1985 Conspecific wounding in free-ranging *R. norvegicus*. *Psychological Record* 35, 329–35

Blanchard R.J. *et al.* 1988 Chronic social stress. *Physiology and Behavior* 43, 1–7

Blanchard R.J. *et al.* 1993 Alcohol, aggression and the stress of subordination. *Journal of Alcohol Studies, Supplement* 11, 146–55

Boice R. 1977 Burrows of wild and albino rats. *Journal of Comparative and Physiological Psychology* 91, 649–61.

Brazelton T.B. 1972 Implications of infant development among the Mayan Indians. *Human Development* 15, 90–111

Booth D.A. 1985 Food-conditioned eating preferences and aversions. In: *Experimental Assessments of Conditioned Food Aversions*, ed. N.S. Braveman & P. Bronstein, New York: Academy of Sciences, pp. 22–41

Buckle A.P 1999 Rodenticides in tropical agriculture. In: *Ecologically-Based Management of Rodent Pests*, ed. G.R. Singleton *et al.* Canberra: ACIAR, pp. 163–77

Calhoun J.B. 1948 Mortality and movement of brown rats. *Journal of Wildlife Management* 12, 167–71

Calhoun J.B. 1949 A method for self-control of population growth among mammals living in the wild. *Science, N.Y.* 109, 333–5

Chitty D. (ed.) 1954 *The Control of Rats and Mice*, vols 1 & 2, Oxford: Clarendon

Chitty D. 1960 Population processes in the vole. *Canadian Journal of Zoology* 38, 99–113

Chopra G. & Sood M.L. 1984 New object and new place reactions of *Rattus meltada*. *Acta Theriologica* 29, 403–12

Cooper R.M. & Zubeck J.P. 1958 Effects of enriched and restricted environments on learning ability. *Canadian Journal of Psychology* 12, 159–64

Cowan P.E. 1983 Exploration in small mammals. In: *Exploration in Animals and Humans,* ed. J. Archer & L. Birke, Wokingham: van Nostrand, pp. 147–75

Davis C.M. 1939 The self-selection of diets by young children. *Canadian Medical Association Journal* 41, 257–61

Davis D.E. 1953 The characteristics of rat populations. *Quarterly Review of Biology* 28, 373–401

References

Defoe D. 1969 (first published 1772) *A Journal of the Plague Year*, ed. L. Landa, London: Oxford University Press

Dickinson A. 1980 *Contemporary Animal Learning Theory*, Cambridge: Cambridge University Press

Dickman C.R. 1999 Rodent-ecosystem relationships. In: *Ecologically-Based Management of Rodent Pests*, ed. G.R. Singleton *et al.*, Canberra: ACIAR, pp. 113–33

Doty R.E. 1938 The prebaited feeding-station method of rat control. *Hawaiian Planters Record* 42, 39–76

Doty R.L. 1985 Humans. In: *Social Odours in Mammals*, vol. 2, ed. R.E. Brown & D.W. Macdonald, Oxford: Clarendon, pp. 805–32

Douglas M. 1966 *Purity and Danger*, London: Routledge & Kegan Paul

Elton C.S. 1942 *Voles, Mice and Lemmings*, Oxford: Clarendon

Emlen J.T., Stokes A.F. & Winser C.P. 1948 The rate of recovery of decimated populations of brown rats. *Ecology* 29, 133–45

Ellerman J.R. 1941 *Families and Genera of Living Rodents*, London: British Museum

Ewer R.F. 1972 The biology and behaviour of *Rattus rattus*. *Animal Behaviour Monographs* 4, 125–74

Galef B.G. 1988 Communication of information concerning distant diets. In: *Social Learning*, ed. T.R. Zentall & B.G. Galef, Hillsdale, N.J.: Erlbaum, pp. 119–39

Galef B.G. 1992 The question of animal culture. *Human Nature* 3, 157–78

Galef B.G. 1996 Social enhancement of food preferences. In: *Social Learning in Animals*, ed. C.M. Heyes & B.G. Galef, San Diego: Academic Press, pp. 49–64

Garcia J. *et al.* 1985 A general theory of aversion learning. In: *Experimental Assessments of Conditioned Food Aversions*, ed. N.S. Braveman & P. Bronstein, New York: Academy of Sciences, pp. 8–21

Gratz N.G. 1988 Rodents and human disease. In: *Rodent Pest Management*, ed. I. Prakash, Boca Raton: CRC Press, pp. 101–69

Harris L.J. *et al.* 1933 Appetite and choice of diet. *Proceedings of the Royal Society B*, 113, 161–90

Hebb D.O. 1949 *The Organization of Behavior*, London: Chapman & Hall

Hecker J.E.C. 1844 *Epidemics of the Middle Ages*. London: Sydenham Society

Hepper P.G. 1988 Adaptive fetal learning. *Animal Behaviour* 36, 935–6

Heyes C.M. 1993 Imitation, culture and cognition. *Animal Behaviour* 46, 999–1010

Hinde R.A. 1956 Ethological models and the concept of 'drive'. *British Journal of the Philosophy of Science* 6, 321–31

Holst D.v. 1986 Psychosocial stress in its pathophysiological effects in tree shrews. In: *Biological and Psychological Factors in Cardiovascular Disease*, ed. T.H. Schmidt, pp. 476–90

Holst D.v. 1998 The concept of stress and its relevance for animal behavior. *Advances in the Study of Behavior* 27, 1–131

Inglis I.R. *et al.* 1996 Foraging behaviour of wild rats. *Applied Animal Behaviour Science* 47, 175–90

Innes M. 1938 *Lament for a Maker*, London: Gollancz

James W. 1890 *Principles of Psychology*, New York: Holt

Kagan J. 1979 *The Growth of the Child*, Hassocks, Sussex: Harvester

Kalat J.W. 1985 Taste-aversion learning. In: *Issues in the Ecological Study of Learning*, ed. T.D. Johnson & A.T. Pietrewicz, London: Erlbaum, pp. 119–41

Kavanau J.L. 1967 Behavior of captive white-footed mice. *Science, N.Y.* 155, 1623–39

Lehrman D.S. 1953 Problems raised by instinct theories. *Quarterly Review of Biology* 28, 337–65

Lewontin R.C. 2000 *The Triple Helix*, Cambridge, MA: Harvard University Press

Logue A.W. 1985 Food aversion learning. In: *Experimental Assessments of Conditioned Food Aversions*, ed. N.S. Braveman & P. Bronstein, New York: Academy of Sciences pp. 316–29

Lorenz K.Z. 1966 *On Aggression*, London: Methuen

References

McNeill W.H. *Plagues and Peoples*, New York: Doubleday

Medawar P.B. 1957 *The Uniqueness of the Individual*, London, Methuen

Mills J.N. 1999 Rodents in emerging human disease. In: *Ecologically-Based Management of Rodent Pests*, ed. G.R. Singleton *et al.*, Canberra: ACIAR, pp. 134–60

Munn N.L. 1950 *Handbook of Psychological Research on the Rat*, New York: Houghton-Mifflin

Nohl J. 1926 *The Black Death*, London: Allen & Unwin

Orwell G. 1949 *Nineteen Eighty-Four*, London: Secker & Warburg

Parrack D.W. 1969 The loss of food to *Bandicota bengalensis*. *Current Science* 38, 93–4

Pavlov I.P. 1927 *Conditioned Reflexes*, London: Oxford University Press

Popper K.R. 1972 *Objective Knowledge*, Oxford: Clarendon

Rackham J. 1979 The introduction of the black rat into Britain. *Antiquity* 53, 112–20

Renner M.J. & Rosenzweig M.R. 1987 *Enriched and Impoverished Environments*, New York: Springer-Verlag

Revusky S. 1977 Learning as a general process. In: *Food Aversion Learning*, ed. N.W. Milgram *et al.*, New York: Plenum, pp. 1–71

Richter C.P. 1957 The phenomenon of sudden death in animals and man. *Psychosomatic Medicine* 19, 191–8

Ristau C.A. (ed.) 1991 *Cognitive Ethology: the Minds of Other Animals*, Hillsdale, NJ: Erlbaum

Robinson R. 1965 *Genetics of the Norway Rat*, Oxford: Pergamon

Russell B. 1927 *An Outline of Philosophy*, London: Allen & Unwin

Sachser N. & Lick C. 1989 Social stress in guinea pigs. *Physiology & Behavior* 46, 137–44

Saltmarsh J. 1941 Plague and economic decline in England in the later Middle Ages. *Cambridge Historical Journal* 7, 23–41

Save E. *et al.* 1998 Landmark use and the cognitive map in the rat. In: *Spatial Representation in Animals*, ed. S. Healy, Oxford: Oxford University Press, pp. 119–32

Selye H. 1957 *The Stress of Life*, London: Longmans, Green

Serafini-Sauli J.P. 1982 *Giovanni Boccaccio*, Boston: Twayne

Shorten M. 1954 The reaction of the brown rat towards changes in its environment. In: *The Control of Rats and Mice*, vol. 2, ed. D. Chitty, Oxford: Clarendon, pp. 307–34

Singleton G.R. *et al.* (ed.) 1999 *Ecologically-Based Management of Rodent Pests*, Canberra: ACIAR

Skinner B.F. 1938 *The Behavior of Organisms*, New York: Appleton-Century

Skinner B.F. 1987 *Upon Further Reflection*, Englewood-Cliffs, N.J.: Prentice-Hall

Spillett J.J. 1968 *The Ecology of the Lesser Bandicoot Rat*, Bombay: Bombay Natural History Society

Stefanski V. 1998 Social stress in loser rats. *Physiology & Behavior* 63, 605–13

Stevenson M.F. 1983 The captive environment. In: *Exploration in Animals and Humans,* ed. J. Archer & L. Birke, Wokingham: van Nostrand, pp. 176–97

Stratakis C.A. & Chrousos G.P. 1995 The stress system. *Annals of the New York Academy of Sciences* 771, 1–18

Suzuki D.T. *et al.* 1996 *An Introduction to Genetic Analysis*, San Francisco: Freeman

Telle H.-J. 1966 Beitrag zur Kenntniss der Verhaltensweise von Ratten. *Zeitschrift für angewandte Zoologie* 53, 129–96

Terkel J. 1995 Cultural transmission in the black rat. *Advances in the Study of Behavior* 24, 119–54

Tinbergen N. 1951 *The Study of Instinct*, Oxford: Clarendon

Tolman E.C. 1958 *Behavior and Psychological Man*, Berkeley: University of California Press

Wallace R.J. & Barnett S.A. 1990 Avoidance of new objects by the black rat. *International Journal of Comparative Psychology* 3, 253–65

Webster J.P. & Macdonald D.W. 1995 Parasites of wild brown rats. *Parasitology* 111, 247–55

Wilson F.P. 1927 *The Plague in Shakespeare's London*, Oxford: Clarendon

Wittgenstein L. 1958 *Philosophical Investigations*, Oxford: Blackwell

References

Wood B.J. 1971 Investigations of rats in rice fields. *Pest Articles and News Summaries* 17, 180–93

Wood J.G. 1870 *The Natural History of Man*, vol. 2, London: Routledge

Ziegler P. 1969 *The Black Death*, London: Collins

INDEX

Adams N. 144
adrenals 167–9, 171
Agar W.E. 126
aggression 129, 138
Aisner R. 105
alligators 152
allogrooming 131
anthropomorphism 54, 129, 165
anticoagulants 100–101, 158–9
Apollo 6
Atharva Veda 3
attack 133, 134, 135
aversions, learned 98, 100–4

bacillus of plague 39–41
bait shyness 98, 100–4
Bammer G. 62
bandicoot rat 12

Bandicota bengalensis see mole rat
Bandicota indica 12
Beach F.A. 143
Beer C.G. 88
behaviourism 71–7, 79
Bell T. 4
Berdoy M. 101, 111, 137
Berkenhout J. 18
Bernard C. 165
Bernstein I.L. 104
Bhagavat Puran 32
Bible, The 5, 29
Bills of Mortality 39
Black Death *see* plague
black rat 15–16, 136
    and cone stripping 105–6
    and kidney disease 170
    and leptospirosis 26
    and plague 40, 42

and teaching 106
conflict among 142
in India 12
pheromones of 133
wounding by 139
Blanchard R.J. 139, 144
Boccacio G. 34-6
*Bodily Changes in Pain, Hunger,
Fear and Rage* 166
Boice R. 62, 144
Boojum 143
Booth D A 98
boredom 77–8, 85
'boxing' 133, 136
brain, growth of 120–1
Brazelton B. 87
Buckle A.P. 159
Bureau of Animal Population
57
bush rat 62, 134–6, 142

Calhoun J.B. 142, 161
Cannon W.B. 166
Cavia aperea see cavy
cavy 144–5
census 99
centipede baffled 89
Chauliac G. de 33
Chernobyl 154
Chitty D. 55, 57
Chrousos G. 167
classifying 14–16, 18
Clyn J. 34

cognition 66, 81–3, 88–90; *see
also* cognitive ethology,
cognitive map
cognitive ethology 89
cognitive map 82–3; *see also*
cognition
commensalism 61–2, 154
conditional reflex 71, 73–4
Cooper R.M. 118
cotton rat 46
Cowan P.E. 61
CR *see* conditional reflex
crawling under 131, 132
Cricetomys gambianus 25
crowding *see* populations
curiousity *see* neophilia
curses 3–4
curtaneous stimulation 122
cytokines 171

Darwin C. 58, 123; *see also*
natural selection
Davis C.M. 97
Davis D.E. 157
death, unexplained ch. 10,
139–42
Decameron, The 35
Defoe D. 38
density related effects 151–3
Descartes R. 175
deutero-learning 81
dietary self selection 94–5, 97
by people 97–8

diet 94–6
    self selection of 97–8
dogs and plague 32
domestication 21–3, 143–4
dominance 129
dominance hierarchy *see* status
    system
Doty R.E. 54–5
Doty R.L. 112
Douglas M. 5
drive 138, 151; *see also* instinct

Ebola virus 45–6
economic threshold 158
Elton C.S. 48, 57, 160
Emlen J.T. 157
environmentalism *see* heredity
    and environment
epigenesis *see* heredity and
    environment
Ewer R.F. 142
experimental neurosis 74
exploration 68–71, 78; *see also*
    neophilia, sampling
    in residential maze 59
exploratory learning 79
extinction (behavioural) 74

flagellants 29–30
fleas 40–1
folk psychology 88
food ch.6
Fox I.A. 131

Frantz S.C. 162
Frazer J. 4

Galef B.G. 109–11
Ganesa 8–9
Garcia J. 102
genes 115, 123, 125
Genesis 22
gerbil, Indian 11
glomerulonephritis 169–70
glucocorticoids 168
Golden Bough, The 4
Gratz N.G. 24–5
guinea-pig *see* cavy

haemorragic fevers 44–8
Harris L.J. 94
Harvey W. 175
Hawaii 54–5
Hebb D.O. 80–2
Hecker J.F.C. 29
Hepper P. 111
Heraclitus 176
heredity and environment 114,
    118–19
    and maternal behaviour 151
    interaction of 123, 127, 144
Heyes C.M. 109
History of British Quadrupeds, A
    4
hoarding 95–6
Hocking W.E. 131
Holst D, von 140–1, 170

homeostasis 165–70
hypoglycaemia 167

imitation *see* learning, social
immune system 168, 170
infants and stimulation 86
Inglis I.R. 93
Innes M. 7
instinct 94, 96, 117; *see also*
  drive, heredity and
  environment
  maternal 151
intelligence 54, 65, 67; *see also*
  cognition
  obscurity of 90
Irulas 12–13

Jews, persecuted 30
Johnson S. 6, 112
Journal of a Plague Year, A 38

Kagan J. 86, 87
Kavanagh L. 60
kidney disease 169–70
killing 139
Kitasato S. 40

laboratory rats *see* Norway rat,
  domestic
Lamarckism 123–7
Landa L. 39
Larousse Gastronomique 11
Lassa fever 45
latent learning 79

learning ch. 5
  after delay 101–3
  early 111–13
  limitations of 113–14
  social 108–11
*Leçons sur les Phénomènes de la
  Vie* 165
Leptospira interrogans 25
leptospirosis 24–7
Lewontin R.C. 164, 175
life, characteristics of 174
longhaired rat 62, 133, 170
  females of 142
Lorenz K.Z. 129

Macdonald D.W. 47
magic 8–9, 30–1, 165
maze
  and learning 65–6
  early design of 21
  learning, genetics of 116–19
  residential 59, 69
  students in 67–8
  Y- 69, 70
McDougall W.M. 126
mechanical psychology 71–7
Medawar P.B. 124
meals 93–4
Mills J.W. 44
minds of animals *see* cognition
mole rat 11, 16–17, 94
  and leptospirosis 26
  and neophobia 61

hoarding by 95–6
populations of 162
mud fever 24–7
Munn N.L. 143
murine typhus 44
mutation 125

natural selection 116; *see also*
Darwin C.
and neophobia 61
neophilia 58–62, 68–71, 77–8
of children 86–7
neophobia 56–8
and aversions 98, 100
flavor 58
new object reaction *see*
neophobia
Nohl J. 19, 30
'Norway' rat 14–19; *see also*
Norway rat, domestic
and exploration 59, 68–71
and neophobia 56–8
as plague carrier 42
as virus carrier 47
breeding of 150
domestication of 16, 21–3
mating by 137
predation by 91
Norway rat, domestic 21–3,
143–4
and feeding 106–11, 120
gentled 119–21
nutrition 91–6

Ockham's razor 79
odors
as signals *see* pheromones
repellent 58
Orwell G. 6
Oxford zoologists 55, 57

Parrack D.W. 95
patrolling 68–71
Pavlov I.P. 71–4, 84, 167
Pepys S. 38
Petty W. 39
pheromones 133
plague 27–44
play 79–80, 136
plus-maze *see* maze, residential
poison shyness 98
Popper K.R. 172, 173
populations 149–63
prebaiting 100
predation 91
public health 36–9

ratcatchers 32
rats *see* under names of species
Rattus argentiventer 158
Rattus exulans 139
Ruttus fuscipes see bush rat
Rattus lutreolus 62, 134
Rattus meltada 11, 61
Rattus norvegicus see 'Norway'
rat
Rattus rattus see black rat

Rattus villosissimus see
  longhaired rat
Renner M.J. 121
Revusky S. 89, 102–3
rice rat 158
Richter C.P. 165
Rosenzweig M.R. 121
Russell B. 84

Sacher N. 144
Saltmarsh J. 34
sampling 69, 95
Sandford M.H.R. 170
Save E. 83
scent marking *see* pheromones
schedule of reinforcement 75
scurvy 10
Selye H. 166
Serafini-Sauli J. 35
sewers 152, 154–7
Shaw G.B. 121
shock disease 164
Shorten M. 54
Sigmodon 46
signals 130–6
Simon A. 11
Skinner B.F. 72–9, 85, 88–9
Skinner box 75–7, 93
Small W.S. 21
Smart J.L. 120
Smith P. 101
Smith R. 54
social feeding 105–13
social learning 108

social sedation 131, 159–61
social signals 130–6, 142, 145
social stress ch. 10, 139–45,
  160–3
  favourable aspects of 166–7
sounds as signals 134
Spillett J.J. 162
spontaneous alternation 70
status system 129–30, 137,
  139
stimulation and intelligence
  77–81, 85–8
Stratakis C. 167
stress
  see social stress
Stress of Life, The 166
students
  biased 119–21
  in mazes 67–8
subordinacy 129
surplus baiting 99
swamp rat 62, 142
synapse 120

taming 16, 21
Tatera indica 11
teaching 108–9
Telle H.-J. 142
Terkel J. 105–6
territory 129, 130, 161–2
thinking *see* cognition
Thorndike E.L. 62
threat posture (TP) 131–4
Tinbergen N. 129–30

Tolman E.C. 65, 116
tree shrew 140–1, 170–1
trial and error 76
Tryon R.C. 117
Tupaia belangeri see tree shrew

Ulrich R.E. 85

vibrissae 66
viruses 44–9
voodoo death 165

Wallace R.J. 61
warfarin 100–1, 158
Webster J.P. 47
Weil's disease 24–7
Whitaker R. 12

Widdowson E.M. 120
Wilson F.P. 31
Wittgenstein L. 89
Wood B.J. 158
Wood J.G. 10
World Health Organization 49

Xenopsylla cheopis 40

Yersin G.A.E. 40
Yersinia pestis 39–41

Ziegler P. 29
zoomophism 84
Zubeck J.P. 118